Data Science: The Hard Parts
Techniques for Excelling at Data Science

Daniel Vaughan

Beijing · Boston · Farnham · Sebastopol · Tokyo

Data Science: The Hard Parts

by Daniel Vaughan

Published by O'Reilly Media, Inc., 1005 Gravenstein Highway North, Sebastopol, CA 95472.

O'Reilly books may be purchased for educational, business, or sales promotional use. Online editions are also available for most titles (*http://oreilly.com*). For more information, contact our corporate/institutional sales department: 800-998-9938 or *corporate@oreilly.com*.

Acquisitions Editor: Aaron Black	**Indexer:** nSight, Inc.
Development Editor: Corbin Collins	**Interior Designer:** David Futato
Production Editor: Jonathon Owen	**Cover Designer:** Karen Montgomery
Copyeditor: Sonia Saruba	**Illustrator:** Kate Dullea
Proofreader: Piper Editorial Consulting, LLC	

November 2023:   First Edition

Revision History for the First Edition

2023-10-31:   First Release

See *http://oreilly.com/catalog/errata.csp?isbn=9781098146474* for release details.

978-1-098-14647-4

[LSI]

This book is dedicated to my brother Nicolas,
whom I love and admire very much.

Table of Contents

Part II. Machine Learning

Preface

I'll posit that learning and practicing data science is *hard*. It is hard because you are expected to be a great programmer who not only knows the intricacies of data structures and their computational complexity but is also well versed in Python and SQL. Statistics and the latest machine learning predictive techniques ought to be a second language to you, and naturally you need to be able to apply all of these to solve actual business problems that may arise. But the job is also hard because you have to be a great communicator who tells compelling stories to nontechnical stakeholders who may not be used to making decisions in a data-driven way.

So let's be honest: it's almost self-evident that the theory and practice of data science is *hard*. And any book that aims at covering the *hard parts* of data science is either encyclopedic and exhaustive, or must go through a preselection process that filters out some topics.

I must acknowledge at the outset that this is a selection of topics that *I* consider the *hard parts* to learn in data science, and that this label is subjective by nature. To make it less so, I'll pose that it's not that they're *harder* to learn because of their complexity, but rather that at this point in time, the profession has put a low enough weight on these as *entry* topics to have a career in data science. So in practice, they are harder to learn because it's hard to find material on them.

The data science curriculum usually emphasizes learning programming and machine learning, what I call the *big themes* in data science. Almost everything else is expected to be learned on the job, and unfortunately, it really matters if you're lucky enough to find a mentor where you land your first or second job. Large tech companies are great because they have an equally large talent density, so many of these somewhat *underground* topics become part of local company subcultures, unavailable to many practitioners.

This book is about techniques that will help you become a more productive data scientist. I've divided it into two parts: Part I treats topics in data analytics and on the softer side of data science, and Part II is all about machine learning (ML).

While it can be read in any order without creating major friction, there are instances of chapters that make references to previous chapters; most of the time you can skip the reference, and the material will remain clear and self-explanatory. References are mostly used to provide a sense of unity across seemingly independent topics.

Part I covers the following topics:

Chapter 1, "So What? Creating Value with Data Science"
What is the role of data science in *creating value* for the organization, and how do you measure it?

Chapter 2, "Metrics Design"
I argue that data scientists are best suited to improve on the *design of actionable metrics*. Here I show you how to do it.

Chapter 3, "Growth Decompositions: Understanding Tailwinds and Headwinds"
Understanding what's going on with the business and coming up with a compelling narrative is a common ask for data scientists. This chapter introduces some *growth decompositions* that can be used to automate part of this workflow.

Chapter 4, "2×2 Designs"
Learning to simplify the world can take you a long way, and *2×2 designs* will help you achieve that, as well as help you improve your communication with your stakeholders.

Chapter 5, "Building Business Cases"
Before starting a project, you should have a *business case*. This chapter shows you how to do it.

Chapter 6, "What's in a Lift?"
As simple as they are, *lifts* can speed up analyses that you might've considered doing with machine learning. I explain lifts in this chapter.

Chapter 7, "Narratives"
Data scientists need to become better at storytelling and structuring compelling *narratives*. Here I show you how.

Chapter 8, "Datavis: Choosing the Right Plot to Deliver a Message"
Investing enough time on your *data visualizations* should also help you with your narrative. This chapter discusses some best practices.

Part II is about ML and covers the following topics:

Chapter 9, "Simulation and Bootstrapping"
Simulation techniques can help you strengthen your understanding of different prediction algorithms. I show you how, along with some caveats of using your

favorite regression and classification techniques. I also discuss *bootstrapping* that can be used to find confidence intervals of some hard-to-compute estimands.

Chapter 10, "Linear Regression: Going Back to Basics"
Having some deep knowledge of *linear regression* is critical to understanding some more advanced topics. In this chapter I go back to basics, hoping to provide a stronger intuitive foundation of machine learning algorithms.

Chapter 11, "Data Leakage"
What is *data leakage*, and how can you identify it and prevent it? This chapter shows how.

Chapter 12, "Productionizing Models"
A model is only good if it reaches the *production stage*. Fortunately, this is a well-understood and structured problem, and I show the most critical of these steps.

Chapter 13, "Storytelling in Machine Learning"
There are some great techniques you can use to open the black box and excel at *storytelling in ML*.

Chapter 14, "From Prediction to Decisions"
We create value from improving our decision-making capabilities through data- and ML-driven processes. Here I show you examples of how to move from *prediction to decision*.

Chapter 15, "Incrementality: The Holy Grail of Data Science?"
Causality has gained some momentum in data science, but it's still considered somewhat of a niche. In this chapter I go through the basics, and provide some examples and code that can be readily applied in your organization.

Chapter 16, "A/B Tests"
A/B tests are the archetypical example of how to estimate the incrementality of alternative courses of action. But experiments require some strong background knowledge of statistics (and the business).

The last chapter (Chapter 17) is quite unique because it's the only one where no techniques are presented. Here I speculate on the future of data science with the advent of generative artificial intelligence (AI). The main takeaway is that I expect the job description to change radically in the next few years, and data scientists ought to be prepared for this (r)evolution.

This book is intended for data scientists of all levels and seniority. To make the most of the book, it's better if you have some medium-to-advanced knowledge of machine learning algorithms, as I don't spend any time introducing linear regression, classification and regression trees, or ensemble learners, such as random forests or gradient boosting machines.

Conventions Used in This Book

The following typographical conventions are used in this book:

Italic
> Indicates new terms, URLs, email addresses, filenames, and file extensions.

`Constant width`
> Used for program listings, as well as within paragraphs to refer to program elements such as variable or function names, databases, data types, environment variables, statements, and keywords.

 This element signifies a tip or suggestion.

 This element signifies a general note.

 This element indicates a warning or caution.

Using Code Examples

Supplemental material (code examples, exercises, etc.) is available for download at *https://oreil.ly/dshp-repo*.

If you have a technical question or a problem using the code examples, please send email to *bookquestions@oreilly.com*.

This book is here to help you get your job done. In general, if example code is offered with this book, you may use it in your programs and documentation. You do not need to contact us for permission unless you're reproducing a significant portion of the code. For example, writing a program that uses several chunks of code from this book does not require permission. Selling or distributing examples from O'Reilly books does require permission. Answering a question by citing this book and quoting example code does not require permission. Incorporating a significant amount of example code from this book into your product's documentation does require permission.

We appreciate, but generally do not require, attribution. An attribution usually includes the title, author, publisher, and ISBN. For example: "*Data Science: The Hard Parts* by Daniel Vaughan (O'Reilly). Copyright 2024 Daniel Vaughan, 978-1-098-14647-4."

If you feel your use of code examples falls outside fair use or the permission given above, feel free to contact us at *permissions@oreilly.com*.

O'Reilly Online Learning

 For more than 40 years, *O'Reilly Media* has provided technology and business training, knowledge, and insight to help companies succeed.

Our unique network of experts and innovators share their knowledge and expertise through books, articles, and our online learning platform. O'Reilly's online learning platform gives you on-demand access to live training courses, in-depth learning paths, interactive coding environments, and a vast collection of text and video from O'Reilly and 200+ other publishers. For more information, visit *https://oreilly.com*.

How to Contact Us

Please address comments and questions concerning this book to the publisher:

O'Reilly Media, Inc.
1005 Gravenstein Highway North
Sebastopol, CA 95472
800-889-8969 (in the United States or Canada)
707-829-7019 (international or local)
707-829-0104 (fax)
support@oreilly.com
https://www.oreilly.com/about/contact.html

We have a web page for this book, where we list errata, examples, and any additional information. You can access this page at *https://oreil.ly/data-science-the-hard-parts*.

For news and information about our books and courses, visit *https://oreilly.com*.

Find us on LinkedIn: *https://linkedin.com/company/oreilly-media*

Follow us on Twitter: *https://twitter.com/oreillymedia*

Watch us on YouTube: *https://youtube.com/oreillymedia*

Acknowledgments

I presented many of the topics covered in the book at Clip's internal technical seminars. As such I'm indebted to the amazing data team that I had the honor of leading, mentoring, and learning from. Their expertise and knowledge have been instrumental in shaping the content and form of this book.

I'm also deeply indebted to my editor, Corbin Collins, who patiently and graciously proofread the manuscript, found mistakes and omissions, and made suggestions that radically improved the presentation in many ways. I would also like to express my sincere appreciation to Jonathon Owen (production editor) and Sonia Saruba (copyeditor) for their keen eye and exceptional skills and dedication. Their combined efforts have significantly contributed to the quality of this book, and for that, I am forever thankful.

Big thanks to the technical reviewers who found mistakes and typos in the contents and accompanying code of the book, and who also made suggestions to improve the presentation. Special thanks to Naveen Krishnaraj, Brett Holleman, and Chandra Shukla for providing detailed feedback. Many times we did not agree, but their constructive criticism was at the same time humbling and reinforcing. Needless to say, all remaining errors are my own.

They will never read this, but I'm forever grateful to my dogs, Matilda and Domingo, for their infinite capacity to provide love, laughter, tenderness, and companionship.

I am also grateful to my friends and family for their unconditional support and encouragement. A very special thank-you to Claudia: your loving patience when I kept discussing some of these ideas over and over, even when they made little to no sense to you, cannot be overstated.

Finally, I would like to acknowledge the countless researchers and practitioners in data science whose work has inspired and informed my own. This book wouldn't exist without their dedication and contributions, and I am honored to be a part of this vibrant community.

Thank you all for your support.

PART I
Data Analytics Techniques

## CHAPTER 1

So What? Creating Value with Data Science

Data science (DS) has seen impressive growth in the past two decades, going from a relatively niche field that only the top tech companies in Silicon Valley could afford to have, to being present in many organizations across many sectors and countries. Nonetheless, many teams still struggle with generating measurable value for their companies.

So what is the value of DS to an organization? I've found that data scientists of all seniorities struggle with this question, so it's no wonder the organizations themselves do so. My aim in this first chapter is to delineate some basic principles of value creation with DS. I believe that understanding and internalizing these principles can help you become a better data scientist.

What Is Value?

Companies exist to create value to shareholders, customers, and employees (and hopefully society as a whole). Naturally, shareholders expect to gain a return on their investment, relative to other alternatives. Customers derive value from the consumption of the product, and expect this to be at least as large as the price they paid.

In principle, all teams and functions ought to contribute in some measurable way to the process of value creation, but in many cases quantifying this is far from obvious. DS is not foreign to this lack of measurability.

In my book *Analytical Skills for AI and Data Science* (O'Reilly), I presented this general approach to value creation with data (Figure 1-1). The idea is simple: data by itself creates no value. The value is derived from the quality of the decisions that are made with it. At a first level, you *describe* the current and past state of the company. This is usually done with traditional business intelligence (BI) tools such as dashboards and reports. With machine learning (ML), you can make *predictions* about the future state and attempt to circumvent the uncertainty that makes the decision process considerably harder. The summit is reached if you can automate and *optimize* some part of the decision process. That book was all about helping practitioners make better decisions with data, so I will not repeat myself here.

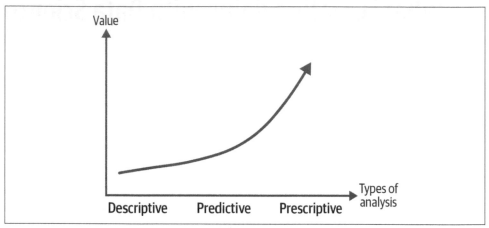

Figure 1-1. Creating value with data

As intuitive as it may be, I've found that this depiction is too general and abstract to be used in practice by data scientists, so over time I've translated this into a framework that will also be handy when I introduce the topic of narratives (Chapter 7).

It boils down to the same principle: incremental value comes from improving an organization's decision-making capabilities. For this, you really need to understand the business problem at hand (*what*), think hard about the levers (*so what*), and be proactive about it (*now what*).

What: Understanding the Business

I always say that a data scientist ought to be as knowledgeable about the business as their stakeholders. And by *business* I mean *everything*, from the operational stuff, like understanding and proposing new metrics (Chapter 2) and levers that their stakeholders can pull to impact them, to the underlying economic and psychological factors that underly the business (e.g., what drives the consumer to purchase your product).

Sounds like a lot to learn for a data scientist, especially since you need to keep updating your knowledge on the ever-evolving technical toolkit. Do you really have to do it? Can't you just specialize on the technical (and fun) part of the algorithms, tech stack, and data, and let the stakeholders specialize on their (less fun) thing?

My first claim is that *the business is* fun! But even if you don't find it exhilarating, if data scientists want to get their voices heard by the actual decision-makers, it is absolutely necessary to gain their stakeholders' respect.

Before moving on, let me emphasize that data scientists are rarely the actual decision-makers on business strategy and tactics: it's the stakeholders, be it marketing, finance, product, sales, or any other team in the company.

How to do this? Here's a list of things that I've found useful:

Attend nontechnical meetings.
No textbook will teach you the nuts and bolts of the business; you really have to be there and learn from the collective knowledge in your organization.

Get a seat with the decision-makers.
Ensure that you're in the meetings where decisions are made. The case I've made for my teams at organizations with clearly defined silos is that it is in the best interest of everyone if they're present. For example, how can you come up with great features for your models if you don't understand the intricacies of the business?

Learn the Key Performance Indicators (KPIs).
Data scientists have one advantage over the rest of the organization: they *own* the data and are constantly asked to calculate and present the key metrics of the team. So you *must* learn the key metrics. Sounds obvious, but many data scientists think this is boring, and since they don't own the metric—in the sense that they're most likely not responsible for attaining a target—they are happy to delegate this to their stakeholders. Moreover, data scientists ought to be experts at metrics design (Chapter 2).

Be curious and open about it.
Data scientists ought to embrace curiosity. By this I mean not being shy about asking questions and challenging the set of accepted facts in the organization. Funny enough, I've found that many data scientists lack this overall sense of curiosity. The good thing is that this can be learned. I'll share some resources at the end of the chapter.

Decentralized structures.
This may not be up to you (or your manager or your manager's manager), but companies where data science is embedded into teams allow for business specialization (and trust and other positive externalities). Decentralized data science structure organizations have teams with people from different backgrounds (data scientists, business analysts, engineers, product, and the like) and are great at making everyone experts on their topic. On the contrary, centralized organizations where a group of "experts" act as consultants to the whole company also have advantages, but gaining the necessary level of business expertise is not one of them.

So What: The Gist of Value Creation in DS

Why is your project important to the company? Why should anyone care about your analysis or model? More importantly, what actions are derived from it? This is at the crux of the problem covered in this chapter, and just in passing I consider it one of those seniority-defining attributes in DS. When interviewing candidates for a position, after the necessary filter questions for the technical stuff, I always jump into the *so what* part.

I've seen this mistake over and over: a data scientist spends a lot of time running their model or analysis, and when it's time to deliver the presentation, they just read the nice graphs and data visualizations they have. Literally.

Don't get me wrong, explaining your figures is super important because stakeholders aren't usually data or data visualization savvy (especially with the more technical stuff; surely they can understand the pie chart on their report). But you shouldn't stop there. Chapter 7 will deal with the practicalities of storytelling, but let me provide some general guidelines on how to develop this skill:

Think about the so what *from the outset.*
Whenever I decide to start a new project, I always solve the problem backwards: how can the decision-maker use the results of my analysis or model? What are the levers that they have? Is it even actionable? Never start without the answers to these questions.

Write it down.
Once you have figured out the *so what*, it's a great practice to write it down. Don't let it play a secondary role by focusing only on the technical stuff. Many times you are so deeply immersed into the technical nitty-gritty that you get lost. If you write it down, the *so what* will act as your North Star in times of despair.

Understand the levers.
The *so what* is all about actionables. The KPIs you care about are generally not directly actionable, so you or someone at the company needs to pull some levers to try to impact these metrics (e.g., pricing, marketing campaigns, sales incentives, and so on). It's critical that you think hard about the set of possible actions. Also, feel free to think out of the box.

Think about your audience.
Do they care about the fancy deep neural network you used in your prediction model, or do they care about how they can use your model to improve their metrics? My guess is the latter: you will be successful if you help them be successful.

Now What: Be a Go-Getter

As mentioned, data scientists are usually not the decision-makers. There's a symbiotic relationship between data scientists and their stakeholders: you need them to put your recommendations into practice, and they need you to improve the business.

The best data scientists I've seen are go-getters who own the project end to end: they ensure that every team plays its part. They develop the necessary stakeholder management and other so-called soft skills to ensure that this happens.

Unfortunately, many data scientists lie on the other side of the spectrum. They think their job starts and ends with the technical part. They have internalized the functional specialization that should be avoided.

Don't be afraid to make product recommendations even when the product manager disagrees with you, or to suggest alternative communication strategies when your marketing stakeholder believes you're trespassing.

That said, be humble. If you don't have the expertise, my best advice before moving to the *now what* arena is to go back to the *what* step and become an expert.

Measuring Value

Your aim is to create measurable value. How do you do that? Here's one trick that applies more generally.

A data scientist does X to impact a metric M with the hope it will improve on the current baseline. You can think of M as a function of X:

$$\text{Impact of } X = M(X) - M(\text{baseline})$$

Let's put this principle into practice with a churn prediction model:

X
    Churn prediction model

M
    Churn rate, i.e., the percentage of active users in period $t - 1$ that are inactive in period t

Baseline
    Segmentation strategy

Notice that *M* is *not* a function of *X*! The churn rate is the same with or without a prediction model. The metric only changes if *you do something with the output of the model*. Do you see how value is derived from actions and not from data or a model? So let's adjust the principle to make it absolutely clear that actions (*A*) affect the metric:

Impact of $X = M(A(X)) - M(A(\text{baseline}))$

What levers are at your disposal? In a typical scenario, you launch a retention campaign targeting only those users with a high probability of becoming inactive the next month. For instance, you can give a discount or launch a communication campaign.

Let's also apply the *what*, *so what*, and *now what* framework:

What
> How is churn measured at your company? Is this the best way to do it? What is the team that owns the metric doing to reduce it (the baseline)? Why are the users becoming inactive? What drives churn? What is the impact on the profit and loss?

So what
> How will the probability score be used? Can you help them find alternative levers to be tested? Are price discounts available? What about a loyalty program?

Now what
> What do you need from anyone at the company involved in the decision-making and operational process? Do you need approval from Legal or Finance? Is Product OK with the proposed change? When is the campaign going live? Is Marketing ready to launch it?

Let me highlight the importance of the *so what* and *now what* parts. You can have a great ML model that is predictive and hopefully interpretable. But if the actions taken by the actual decision-makers don't impact the metric, the value of your team will be zero (*so what*). In a proactive approach, you actually help them come out with alternatives (this is the importance of the *what* and becoming experts on the problem). But you need to ensure this (*now what*). Using my notation, you must own $M(A(X))$, not only *X*.

Once you quantify the incrementality of your model, it's time to translate this to value. Some teams are happy to state that churn decreased by some amount and stop there. But even in these cases I find it useful to come up with a dollar figure. It's easier to get more resources for your team if you can show how much incremental value you've brought to the company.

In the example this can be done in several ways. The simplest one is to be literal about the value.

Let's say that the monthly average revenue per user is R and that the company has base of active users B:

$$\text{Cost of Churn}(A, X) = B \times \text{Churn}(A(X)) \times R$$

If you have 100 users, each one bringing $7 per month, and a monthly churn rate of 10% churn, the company loses $70 per month.

The incremental monetary value is the difference in the costs with and without the model. After factoring out common terms, you get:

$$\Delta\text{Cost of Churn}(A, \text{baseline}, X) = B \times \Delta\text{Churn}(A; X, \text{baseline}) \times R$$

If the previously used segmentation strategy saved $70 per month, and the now laser-focused ML model creates $90 in savings, the incremental value for the organization is $20.

A more sophisticated approach would also include other value-generating changes, for instance, the cost of false positives and false negatives:

False positive
It's common to target users with costly levers, but some of them were never going to churn anyway. You can measure the cost of these levers. For instance, if you give 100 users a 10% discount on the price P, but of these only 95 were actually going to churn, you are giving away $5 \times 0.1 \times P$ in false positives.

False negative
The opportunity cost from having bad predictions is the revenue from those users that end up churning but were not detected by the baseline method. The cost from these can be calculated with the equations we just covered.

Key Takeaways

I will now sum up the main messages from this chapter:

Companies exist to create value. Hence, teams ought to create value.
A data science team that doesn't create value is a luxury for a company. The DS hype bought you some leeway, but to survive you need to ensure that the business case for DS is positive for the company.

Value is created by making decisions.
DS value comes from improving the company's decision-making capabilities through the data-driven, evidence-based toolkit that you know and love.

The gist of value creation is the so what.

Stop at the outset if your model or analysis can't create actionable insights. Think hard about the levers, and become an expert on your business.

Work on your soft skills.

Once you have your model or analysis and have made actionable recommendations, it's time to ensure the end-to-end delivery. Stakeholder management is key, but so is being likeable. If you know your business inside out, don't be shy about your recommendations.

Further Reading

I touch upon several of these topics in my book *Analytical Skills for AI and Data Science* (O'Reilly). Check out the chapters on learning how to ask business questions and finding good levers for your business problem.

On learning curiosity, remember that you were born curious. Children are always asking questions, but as they grow older they forget about it. It could be because they've become self-conscious or a fear of being perceived as ignorant. You need to overcome these psychological barriers. You can check out *A More Beautiful Question: The Power of Inquiry to Spark Breakthrough Ideas* by Waren Berger (Bloomsbury) or several of Richard Feynman's books (try *The Pleasure of Finding Things Out* [Basic Books]).

On developing the necessary social and communication skills, there are plenty of resources and plenty of things to keep learning. I've found *Survival of the Savvy: High-Integrity Political Tactics for Career and Company Success* by Rick Brandon and Marty Seldman (Free Press) quite useful for dealing with company politics in a very pragmatic way.

Extreme Ownership: How U.S. Navy Seals Lead and Win by Jocko Willink and Leif Babin (St. Martin's Press) makes the case that great leaders exercise end-to-end (extreme) ownership.

Never Split the Difference by Chris Voss and Tahl Raz (Harper Business) is great at developing the necessary negotiation skills, and the classic and often-quoted *How to Win Friends and Influence People* by Dale Carnegie (Pocket Books) should help you develop some of the *softer* skills that are critical for success.

## CHAPTER 2

Metrics Design

Let me propose that great data scientists are also great at metrics design. What is metrics design? A short answer is that it is the art and science of finding metrics with good properties. I will discuss some of these desirable properties shortly, but first let me make a case for why data scientists ought to be great at it.

A simple answer is: because if not us, who else? Ideally *everyone* at the organization should excel at metrics design. But data practitioners are the best fit for that task. Data scientists work with metrics all the time: they calculate, report, analyze, and, hopefully, attempt to optimize them. Take A/B testing: the starting point of every good test is having the right output metric. A similar rationale applies for machine learning (ML): getting the correct outcome metric to predict is of utmost importance.

Desirable Properties That Metrics Should Have

Why do companies need metrics? As argued in Chapter 1, good metrics are there to drive actions. With this success criterion in mind, let's reverse engineer the problem and identify necessary conditions for success.

Measurable

Metrics are measurable by definition. Unfortunately, many metrics are imperfect, and learning to identify their pitfalls will take you a long way. So-called *proxy* metrics or *proxies* that are usually correlated to the desired outcome abound, and you need to understand the pros and cons of working with them.[1]

---

1 In linear regression, for example, measurement error on the features creates statistical bias of the parameter estimates.

A simple example is *intentionality*. Suppose you want to understand the drivers for *early churn* (churn of new users). Some of them never actually intended to use the product and were just trying it out. Hence, measuring intentionality would greatly improve your prediction model. Intentionality isn't really measurable, so you need to find proxies, for instance, the time lag between learning about the app and starting to use it. I'd argue that the faster you start using it, the more intent you have.

Another example is the concept of *habit* used by growth practitioners. Users of an app usually finish onboarding, try the product (the aha! moment), and hopefully reach habit. What is good evidence that a user reached this stage? A common proxy is the number of interactions in the first X days since the user first tried it. To me, habit is all about *recurrence*, whatever that means for each user. In this sense, the proxy is at best an early indicator of recurrence.

Actionable

To drive decisions, metrics must be actionable. Unfortunately, many top-line metrics aren't directly actionable. Think of *revenue*: it depends on the user purchasing the product, and that cannot be forced. But if you decompose the metric into submetrics, some good levers may arise, as I'll show in the examples.

Relevance

Is the metric informative for the problem at hand? I call this property *relevance* since it highlights that a metric is only good relative to a specific business question. I could use *informative*, but all metrics are informative of something. Relevance is the property of having *the right* metric for the right problem.

Timeliness

Good metrics drive actions when you need them to. If I learn that I have terminal cancer, my doctors won't be able to do much about it. But if I get checked regularly, they may find an early symptom, thereby opening the menu of treatments at my disposal.

Customer churn is another example. It's commonly measured and reported using a one-month-of-inactivity window: the percentage of users that were active one month and inactive the next month. Unfortunately, this metric can create false positives: some users were just taking a break and didn't churn.

One way to get a more robust metric is to increase the inactivity window from one to three months, say. The longer the time window, the less likely a user is just taking a break. But the new metric has degraded in terms of timeliness: you now have to wait three months to flag a customer who churned, and it might be too late to launch a retention campaign.

Metrics Decomposition

By decomposing a metric you may be able to improve on any of these properties. I will now cover in detail several tricks that will help you achieve this.

Funnel Analytics

Funnels are a sequence of actions that go one after the other. For example, in the earlier habit example, the user first needs to set up their account, try the product, and use it recurrently. Whenever you have funnels, you can use a simple trick to find submetrics. Let me show the trick in abstract first and then provide some concise examples.

Figure 2-1 shows a typical funnel: it is a sequence of stages between an entry point E and the output M (abusing notation; these also represent the corresponding metric). My objective is to improve on M. Internal stages are denoted as s_1, s_2, s_3, and each provides a metric denoted with an m_i correspondingly indexed.

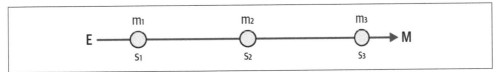

Figure 2-1. A typical funnel

The decomposition works like this: you move from right to left, multiplying by the current submetric, and dividing by the previous one. To guarantee that you never lose the equality, finish by multiplying the metric at the beginning of the funnel (E). Notice that after canceling out common terms, the end result is $M = M$, ensuring that this is indeed a decomposition of the original metric.

$$M = \frac{M}{m_3} \times \frac{m_3}{m_2} \times \frac{m_2}{m_1} \times \frac{m_1}{E} \times E$$

Each fraction can be interpreted as a conversion rate, that is, the percentage of units available in the previous stage that make it to the current stage. Usually one or all of these submetrics have better properties than the original metric M. Now that you understand the technique, it's time to put it into action.

A typical sales funnel works just like this. My aim is to increase sales, but this requires several steps. Here I'll simplify the funnel a bit:

- Lead generation (L: number of leads)
- First contact (C_1: number of first contacts)

- Second contacts (C_2: number of second contacts)
- Make an offer (O: number of offers made)
- Close the sale (S: number of sales)

The decomposition becomes:

$$S = \frac{S}{O} \times \frac{O}{C_2} \times \frac{C_2}{C_1} \times \frac{C_1}{L} \times L$$

To increase the number of sales, you can increase the number of leads, or increase the conversion between stages. Some actions are related to the data scientist (for example, improving the quality of the leads), others to the sales team (such as, whether they are making enough first contacts; if not, the company might need to increase the size of the sales force or hire different people). Maybe they should change the negotiation or pricing strategies to improve the offer-to-sale rate. Or even make improvements on the product! You can have the best leads or the best sales force but still lack product-market fit.

Stock-Flow Decompositions

Stock-flow decompositions are useful when you care about a metric that accumulates. Let's start by defining these concepts: a *stock* variable is one that accumulates and is measured at a specific point in time. *Flow* variables don't accumulate and are measured over a period of time. A useful analogy is a bathtub: the volume of water at time t is equal to the volume at time $t - 1$, plus the water that came in through the faucet between these two moments in time, minus the water that went down the drain.

The most common scenario is when you want to understand Monthly Active Users (*MAU*). I'll spell out the decomposition first, and comment after:

$$\text{MAU}_t = \text{MAU}_{t-1} + \text{Incoming Users}_t - \text{Churned Users}_t$$

If the objective is to grow the *MAU* for the company, you can either increase customer acquisition or reduce churn. *Incoming Users* can potentially be open into *New Users* and *Resurrected Users*, providing at least one new lever.

Similar decompositions apply for any stock variable (such as balances in a bank account).

P×Q-Type Decompositions

Another common scenario is trying to improve revenue. The trick here is to multiply and divide by a *reasonable* metric to arrive at submetrics that are most easily leveraged:

$$\text{Revenue} = \frac{\text{Revenue}}{\text{Units Sold}} \times \text{Units Sold} = \text{Unit Price} \times \text{Sales}$$

This shows how to decompose revenue as the product of unit (average) price and sales: $R = p \times q$. To increase revenue you can either increase the price or sales. Interestingly, sales depend negatively on the price, so the relationship is nonlinear, making it a preferred tool for revenue optimization.

Example: Another Revenue Decomposition

Starting from the fact that revenue is generated by active users, you may attempt a similar decomposition that could prove valuable for certain problems and choices of levers:

$$\text{Revenue} = \frac{\text{Revenue}}{\text{MAU}} \times \text{MAU} = \text{ARPU} \times \text{MAU}$$

I just expressed revenue as a function of Average Revenue per User (ARPU) and active users. I could plug in the MAU stock equation if I want to find even more levers. Similarly, I could also plug in the $p \times q$ decomposition to expand the list.

Example: Marketplaces

As a final example, consider a marketplace: a two-sided platform that matches buyers (B) and sellers (S). Think Amazon, eBay, Uber, Airbnb, and so forth.

Let's consider a simplified funnel:

$$\text{Sellers} \rightarrow \text{Listed Items} \rightarrow \text{Views} \rightarrow \text{Purchases}$$

Under this interpretation, the company first onboards Sellers that start listing items that get viewed and end up in a purchase. Your objective is to increase Purchases.

Using the funnel logic, this translates to (capital letters denote the corresponding metric in each step):

$$P = \frac{P}{V} \times \frac{V}{L} \times \frac{L}{S} \times S$$

To include the other side of the market, let's apply another one of the tricks discussed earlier, so that total Viewed items equals the number of Buyers times the average number of Views per Buyer:

$$V = \frac{V}{B} \times B$$

After some rearranging I arrive at:

$$P = \frac{P}{V} \times \frac{V}{B} \times \frac{L}{S} \times \frac{1}{L} \times B \times S$$

It follows that to increase purchases you can either:

- Increase the checkout efficiency (P/V)
- Increase Buyers' engagement (V/B)
- Increase Sellers' engagement (L/S)
- Increase the volume of Buyers or Sellers

To guarantee that the equality is maintained, I have an extra term that lacks an obvious interpretation ($1/L$). I don't really care about this extra term, since I now have five submetrics that can be leveraged in different ways.[2]

Key Takeaways

These are the key takeaways from this chapter:

You need good metrics to drive actions.
> Metrics design is critical if your aim is to find levers that can drive actions. I have reverse engineered the problem to arrive at some desirable properties for metrics design.

---

2 You can give the extra term a probabilistic interpretation if views are generated at random from the set of listed items. But this really defeats the purpose of the decomposition.

Desirable properties that good metrics should have.
A good metric must be measurable, actionable, relevant, and timely.

Decomposing metrics into submetrics allows you to improve on these properties.
Funnel-type decompositions are easy to use, and once you get used to them, you'll start to see funnels everywhere.

A simple trick of multiplying and dividing by one metric can take you very far. But the choice of that metric is far from obvious, and you need good knowledge of the business to find it.

Metrics design is an iterative process.
It's fine to start with imperfect metrics, but it's even better if you make this a constant iterative process.

Further Reading

You can check out my book *Analytical Skills for AI and Data Science* if you want some complementary information, but this chapter is more comprehensive on the actual techniques used. In that book, I also show how the $R = p \times q$ decomposition can be used for revenue optimization.

A discussion on metrics design for growth enthusiasts can be found in *Hacking Growth: How Today's Fastest-Growing Companies Drive Breakout Success* by Sean Ellis and Morgan Brown (Currency).

While not a book on metrics design but rather on OKRs, *Measure What Matters* by John Doerr (Portfolio) is certainly worth a read. I've used the techniques presented here to find submetrics that can actually be targeted by specific teams. To the best of my knowledge, there's no other published resources on these topics from a data science perspective.

## CHAPTER 3

Growth Decompositions: Understanding Tailwinds and Headwinds

Chapter 2 described some techniques to find better metrics that can drive actions. This chapter deals with a completely different subject: how you can decompose metrics to understand *why* a metric changed. In corporate jargon these changes are usually associated with *tailwinds* or *headwinds*, that is, factors that positively or negatively affect the state of the company.

Why Growth Decompositions?

Data scientists are frequently asked to help understand the root cause of a change in a metric. Why did revenues increase quarter over quarter (QoQ) or month over month (MoM)? In my experience, these are very hard questions to answer, not only because many things can be happening at the same time, but also because some of these underlying causes are not directly measurable or don't provide enough variation to be informative.[1] Typical examples are things like the state of the economy or the regulatory environment, as well as decisions made by competitors.

Nonetheless, I've found that you can use some other source of variations that, when coupled with the following techniques, can give you hints of what's going on.

---

1 In Chapter 10, I discuss why you *need* variation in the inputs to explain variation in the output metric.

Additive Decomposition

As the name suggests, this decomposition is handy when the metric (output) you want to understand can be expressed as the sum of other metrics (inputs). In the case of two inputs, this can be expressed as $y_t = y_{1,t} + y_{2,t}$. Note that I'm using a time subscript.

The decomposition says that the growth of the output from $t - 1$ to t ($g_{y,t}$) is the weighted average of the input's growth rates:

$$g_{y,t} = \omega_{1,t-1} g_{y_1,t} + \omega_{2,t-1} g_{y_2,t}$$

where weights add up to one, $\omega_{1,t-1} + \omega_{2,t-1} = 1$.

Importantly, the weights are the relative importance of each input *in the previous period*. So an input that had a larger share in *t-1* will be given more weight.

Example

The additive setting is quite common in data warehouses where you have fact and dimensional tables. I've found that a grammar analogy helps distinguish one from the other: facts reflect actions or verbs, and dimensions are adverbs that describe the action. Fact tables usually store metrics that are relevant to the company, and dimensional tables store dimensions that help you understand the metrics.

Here's a typical SQL query that generates the dataset needed as input:

```
SELECT DATE_TRUNC('MONTH', ft.fact_timestamp) AS month,
       dt.dimension1 AS dim_values,
       SUM(ft.my_metric) AS monthly_metric
FROM my_fact_table ft
LEFT JOIN my_dim_table dt ON ft.primary_key = dt.primary_key
GROUP BY 1,2
ORDER BY 1,2
```

For example, it can be that the metric is customer purchases and you want to open this by geographical regions. Since total sales must be the sum of sales across regions, this decomposition becomes handy. It will help you understand if growth rates in one or several regions are the main drivers for an acceleration or deceleration at the national level.

The sample query highlights how you can easily create an aggregate table that splits a metric using *different* dimensions. The process looks like this:

1. Create a pipeline that periodically updates the aggregate table across different dimensions.

2. Write a script that computes the decomposition for one dimension and outputs the results as a table (see the GitHub repo (*https://oreil.ly/dshp-repo*)).

3. Loop over all dimensions with that script.

4. The end result is a table with all sources of variations.

At this point, you need your knowledge of the business to identify patterns in the changes. This is usually the hardest part and requires extensive knowledge of the business.

Interpretation and Use Cases

As mentioned, with the additive decomposition, the growth rate in the output equals the weighted average of the growth rates in the inputs. Nonetheless, I prefer to think of growth contributions by each segment or dimensional value, where each contribution is equal to the product of the lagged weight and the corresponding growth rate.

Simplifying the Additive Decomposition

Growth in Output = SUM(inputs' contributions to growth)

The decomposition is especially useful when you have several dimensions that can be used simultaneously and can jointly provide hints about the underlying factors.

Going back to the sales example, you can apply the decomposition using geographical regions, socioeconomic status (SES) of the store's neighborhood, and some type of customer segment (e.g., by tenure).

The conclusion might be something like: national sales decreased 7 percentage points (pp) MoM mainly because:

- The Southwest region decelerated 14 pp MoM.
- Stores in high SES areas decelerated faster.
- Deceleration was relatively uniform across tenures.

As warned earlier, notice that you are *not* really finding the root causes; at best you have enough hints for what's driving the change. Is the Southwest region economy decelerating? Were there changes in pricing in these stores? What is customer satisfaction looking like in high SES customers?

Figure 3-1 shows a waterfall plot of the regional contributions for a simulated example. In this case, there was a 4.6% decline at the national level, explained mainly by a strong deceleration in the Northwest region (5.8 pp). Southwest and West regions also decelerated, and the South had a strong quarter.

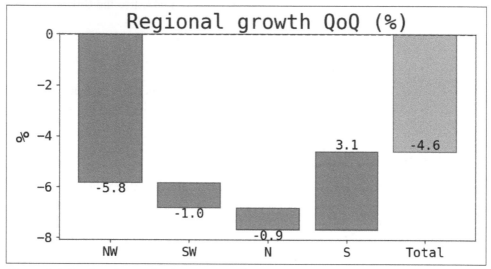

Figure 3-1. Regional contributions to growth

Multiplicative Decomposition

The *multiplicative decomposition* works when the output metric can be expressed as the product of two or more inputs. Chapter 2 showed how these arise naturally in many setups, for instance in $p \times q$ cases.

The decomposition says that whenever $y_t = y_{1,t} \times y_{2,t}$, then:

$$g_{y,t} = g_{1,t} + g_{2,t} + g_{1,t} \times g_{2,t}$$

In words, the growth rate in the output is equal to the sum of the growth rates *and* a combined effect.

Example

Let's use the revenues decomposition from Chapter 2 and see that these are the product of Average Revenue per User (ARPU) and Monthly Active Users (MAU):

Revenues = ARPU × MAU

If revenues grew, it could be because ARPU accelerated, MAU increased, or both changed in the same direction. More importantly, with the decomposition you can actually *quantify* each of these.

Figure 3-2 shows one possible visualization of the decomposition for a simulated ARPU example. In this case, the main driver for MoM growth is a considerable acceleration in average revenue per user (contributing ~31pp or around 96% of the total revenue growth). Notice that the combined effect is very small, as it is the product of the growth rates of the inputs. Many times you can just drop it, if it really is negligible.[2]

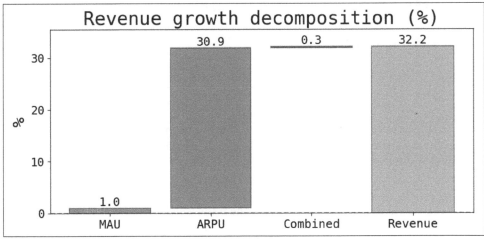

Figure 3-2. Multiplicative decomposition of ARPU

Interpretation

In a multiplicative setting, the growth of the output is the sum of growth in inputs plus a combined effect. With more than two inputs, this still holds but you need to add the sum of the combined effects.

---

2 If you use a log transformation, you can use a Taylor expansion to get the same result that the growth rate of a product is just the sum of the growth rates of the inputs.

> ## Simplifying the Multiplicative Decomposition
> Growth in Output = SUM(growth in inputs) + combined effect

Mix-Rate Decompositions

Mix-rate decompositions take a bit from each of the additive and multiplicative decompositions. Suppose that your output metric is a weighted average of other metrics:

$$y_t = \sum_s w_{s,t} x_{s,t} = \mathbf{w_t} \cdot \mathbf{x_t}$$

where the last equality is just expressing the sum as a dot or inner product of the corresponding vectors (in bold).

Let me spell out the decomposition and then explain the terms:

$$\Delta y_t = \Delta_y^x + \Delta_y^w + \Delta\mathbf{w} \cdot \Delta\mathbf{x}$$

Where:

Δy_t

> First difference for the output metric. I've found that keeping everything as differences—instead of growth rates—is usually all that's needed, and it considerably simplifies the notation.

Δ_y^x

> What would the change in the output be if the weights are kept fixed at the initial values, and *only* the inputs change? The notation signals that only inputs (superscript) are allowed to change the output (subscript).

Δ_y^w

> What would the change in the output be if the inputs are kept fixed at the initial values, and *only* the weights change?

$\Delta\mathbf{w} \cdot \Delta\mathbf{x}$
> This is the inner product of the changes in weights and inputs.

When I first started thinking about this decomposition, I started with the intuition in the second and third points, which are counterfactual (i.e., you can't observe them) and quite useful for storytelling purposes. The math didn't add up, so I had to go through the derivation. I once presented this to a stakeholder and they called it *mix-rate*; it appears that this term was used some time ago, but after searching the web I couldn't find much, so I'm not really sure about its origin or usage. The term is good, though, as there are two potential sources of change:

- Changes in the weights (mix)
- Changes in the inputs (rates)

Example

Weighted averages arise everywhere. Think about this: you have one metric and customer segments. It's intuitive to believe that the metric will be a weighted average of the metrics for the segments. This is always the case with ratio metrics. Let's try it with the average revenue per user for two segments:

$$
\begin{aligned}
ARPU &= \frac{R}{MAU} \\
&= \frac{R_1 + R_2}{MAU_1 + MAU_2} \\
&= \frac{R_1}{MAU_1} \frac{MAU_1}{MAU_1 + MAU_2} + \frac{R_2}{MAU_2} \frac{MAU_2}{MAU_1 + MAU_2} \\
&= \omega_1 ARPU_1 + \omega_2 ARPU_2
\end{aligned}
$$

Note that weights are the relative share of monthly active users in the period for each segment. As usual, weights must add up to one.

Figure 3-3 shows one possible visualization of this decomposition for a simulated dataset for the ARPU example (with three segments). Had there not been any changes in the shares, ARPU would have increased by $3.2 (*rate*); similarly, had there not been any changes in ARPU per segment, average revenue per user would have fallen by $1.6 (*mix*).

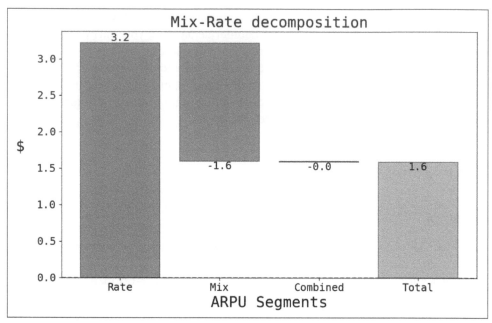

Figure 3-3. Example of a mix-rate decomposition

Interpretation

The interpretation is straightforward: a change in the metric is equal to the sum of the partialled-out parts (that is, fixing one component at initial values and allowing the other to change) and the combined effect of both changes.

> ### Simplifying the Mix-Rate Decomposition
>
> Growth in Metric = SUM(partialled-out effects) + combined effect

As mentioned earlier, I find the first part quite compelling for storytelling purposes, since you're effectively simulating what would've happened if only the weights or the rates had changed.

Mathematical Derivations

Let's dive into the math; understanding the derivation is crucial for coding purposes. I've found myself debugging a function because I did *not* use the right weights, or because the time subscripts were wrong.

In what follows I'll simplify by assuming only two summands (additive), multiples (multiplicative), or segments (mix-rate). It's easy to check that these generalize to more inputs or segments (but you need to be careful, as you can see in the code repo (*https://oreil.ly/dshp-repo*)).

Also, I denote the growth rate of x as $g_t = \dfrac{\Delta x_t}{x_{t-1}}$, with $\Delta x_t := x_t - x_{t-1}$ the first difference of x.

Additive Decomposition

Since y is additive:

$$y_t = y_{1,t} + y_{2,t}$$

Let's take the first differences now to get:

$$\Delta y_t = \Delta y_{1,t} + \Delta y_{2,t}$$

Finally, to get growth rates:

$$\frac{\Delta y_t}{y_{t-1}} = \frac{\Delta y_{1,t}}{y_{1,t-1}} \frac{y_{1,t-1}}{y_{t-1}} + \frac{\Delta y_{2,t}}{y_{2,t-1}} \frac{y_{2,t-1}}{y_{t-1}} = \omega_{1,t-1} \frac{\Delta y_{1,t}}{y_{1,t-1}} + \omega_{2,t-1} \frac{\Delta y_{2,t}}{y_{2,t-1}}$$

or

$$g_{y,t} = \omega_{1,t-1} g_{1,t} + \omega_{2,t-1} g_{2,t}$$

Multiplicative Decomposition

Since y is multiplicative:

$$y_t = y_{1,t} \times y_{2,t}$$

Taking a first difference of the output and adding and subtracting an extra term (that helps factor out extra terms):

$$\Delta y_t = y_{1,t} y_{2,t} - y_{1,t-1} y_{2,t-1} + y_{1,t} y_{2,t-1} - y_{1,t} y_{2,t-1} = y_{1,t} \Delta y_{2,t} + y_{2,t-1} \Delta y_{1,t}$$

To get growth rates, you just need to be a bit careful and remember that the output is multiplicative for all time periods:

$$\frac{\Delta y_t}{y_{t-1}} = \frac{y_{1,t}}{y_{1,t-1}}\frac{\Delta y_{2,t}}{y_{2,t-1}} + \frac{y_{2,t-1}}{y_{2,t-1}}\frac{\Delta y_{1,t}}{y_{1,t-1}} = \left(1 + g_{1,t}\right)g_{2,t} + g_{1,t} = g_{1,t} + g_{2,t} + g_{1,t}g_{2,t}$$

Note that if you have more than two inputs, you need to sum *all* combinations of products.

Mix-Rate Decomposition

Recall that for the mix-rate case, the output metric can be expressed as a weighted average of the metric for segments:

$$y_t = \mathbf{w_t} \cdot \mathbf{x_t}$$

where the weights add up to one, and bold letters denote vectors.

In this case I'll work backward and show that after some simplifications you will arrive at the original expression. Not the most elegant way, but I would rather do it this way instead of adding and subtracting terms that you wonder where they came from.

$$
\begin{aligned}
\Delta_y^x + \Delta_y^w + \Delta\mathbf{w} \cdot \Delta\mathbf{x} &= \underbrace{\mathbf{w_{t-1}} \cdot \Delta\mathbf{x} + \mathbf{x_{t-1}} \cdot \Delta\mathbf{w} + \Delta\mathbf{w} \cdot \Delta\mathbf{x}}_{\text{Replacing the definitions}} \\
&= \underbrace{\Delta\mathbf{x} \cdot \left(\Delta\mathbf{w} + \mathbf{w_{t-1}}\right) + \Delta\mathbf{w} \cdot \mathbf{x_{t-1}}}_{\text{Factoring out } \Delta x} \\
&= \underbrace{\mathbf{x_t}\mathbf{w_t} - \mathbf{x_{t-1}}\mathbf{w_{t-1}}}_{\text{Simplifying}} \\
&= \Delta y_t
\end{aligned}
$$

Key Takeaways

These are the key takeaways from this chapter:

Finding root causes for changes in time is usually very hard.
    You need enough variation in the drivers to estimate impacts.

Growth decompositions are useful to get hints about the underlying root causes.

By exploiting these extra sources of variations (from other input metrics), you're able to hypothesize what drove the change. I've shown three decompositions that might work for the problem you face: additive, multiplicative, and mix-rate.

Further Reading

To the best of my knowledge, there's not much published literature on this. My impression is that this knowledge is shared within company data teams and cultures, but never gets out to the more general public. I learned about the additive decomposition in a previous job, and worked out the other two as needed.

The math is relatively straightforward, so there's no need to develop it further. If you're still interested, the methods I've used can be found in any introductory book or lecture notes on discrete calculus.

## CHAPTER 4

2×2 Designs

Some years ago, when I was starting my career in data science, a consulting firm came to the office and started sketching these extremely simplified views of our business. My immediate reaction was to dismiss these sketches as a trick in their sales-driven bag. Today I embrace them for communication and storytelling purposes, as well as useful aids to simplify a complex business.

I believe that a natural growth path in data science (DS) is to go from making things overly complex to doing *smart* simplification. By *smart* I mean what Einstein expressed when saying you should aim at making "everything as simple as possible, but not simpler." The beauty of this quote is that it shows how difficult it is to achieve this. In this chapter, I'll make a case for using a tool designed for the specific purpose of simplifying a complex world.

The Case for Simplification

You may find it ironic that I make a case for simplification in the age of big data, computational power, and sophisticated predictive algorithms. These tools allow you to navigate the ever-increasing volumes of data and thus have undoubtedly improved data scientists' productivity, but they don't really simplify the world or the business.

Let's stop for a second on this last thought: *if* more data means more complexity, then data scientists are now definitely capable of making sense of more complexity. Nonetheless, the fact that you can make *projections* of high-dimensional data onto lower-dimensional scores does not mean you have a better understanding of how things work.

There are many cases one can make for simplification, from the aesthetic to the more functional and pragmatic. For the data scientist, simplification helps their understanding and framing of what's most important when starting a project. Moreover, it's a great communication tool. As Richard Feynman said, "If you cannot explain something in simple terms, you don't understand it." On the technical side, it's quite common to apply Occam's razor to choose the simplest model that has a given predictive performance.

What's a 2×2 Design?

Figure 4-1 shows a typical *design*. As this last word suggests, you play an active role in deciding which features to concentrate on, which of course vary, depending on the use case.

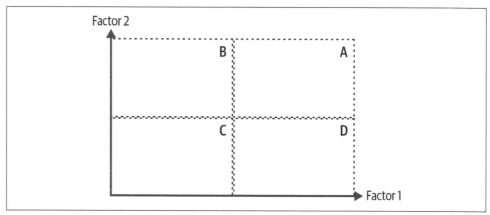

Figure 4-1. A typical 2×2 design

Note how I've simplified the world by concentrating only on two factors or features that I believe are relevant for the task at hand. Factors 1 and 2 vary across the horizontal and vertical axis, respectively. Moreover, I've discretized a possibly continuous world by setting some threshold levels that are represented by dashed vertical and horizontal lines, dividing the world into four quadrants:

A
> Users with *high* factors 1 and 2

B
> Users with *low* factor 1 and *high* factor 2

C
> Users with *low* factors 1 and 2

D
> Users with *high* factor 1 and *low* factor 2

Depending on the use case, I can play around with these thresholds.

In experimental designs, these factors usually correspond to different treatments in the test, such as the color and message used in a banner, or price and frequency of communication. The first example deals with discrete factors, and the latter with continuous features. Needless to say, with discrete factors you lose the sense of ordering explicit in the diagram.

Ideally, every other relevant factor should *remain constant*. This more general scientific principle allows you to single out the impact of these two factors on the metric of interest. In Chapter 10 I will come back to this line of reasoning, but for now note that this *partialling out* is crucial in your attempt to simplify the world: by changing one factor at a time, with everything else fixed, you can gain *some* insight into each factor's role.

In statistical 2×2 designs, this partialling out is guaranteed by using a proper randomization scheme that makes participants in each treatment and control *ex ante equal* on average. This somewhat cryptic phrase means that before the test, treatment and control groups don't differ too much, on average.

These designs are well known by statistical practitioners, and the topic is usually covered when studying analysis of variance (ANOVA). The objective here is to see if there are differences in the means of an outcome metric across groups. Treatments are often discrete, but the design allows for continuous treatments by conveniently setting thresholds.

This same setting can be used in nonexperimental scenarios. The typical example used by consulting firms is to segment the customer base using only two features that may or may not be behavioral. I've commonly used it when I can decompose a metric in a multiplicative way (like the $p \times q$ decomposition seen in Chapter 2).

For instance, take unit price and transactionality. Quadrant A represents customers that are willing to pay high unit prices and transact a lot (yielding high average revenue per user). Note that here I cannot guarantee that *everything else remains constant*, as in the experimental setting. Nonetheless, it still allows me to focus on, and only on, the two features that I care about.

I'll now show some examples.

Example: Test a Model and a New Feature

A typical scenario where I use the 2×2 framework is when I want to simultaneously test a new model and the effectiveness of a lever. Testing the lever is commonly done *without* this framework, just by having two randomized groups: one receives the baseline (control), and the other gets the new lever (treatment). When the experiment is finished, I run the typical statistical testing suite on the differences in means. The 2×2 design expands on this idea by allowing you to also test the performance of your model.

Figure 4-2 shows the 2×2 design. On the horizontal axis I have the probability score (in this example, coming from a classification model). The vertical axis shows whether I've turned on or off the lever considered for the test: lever *on* means that you display the new alternative to some users, and *off* means that the baseline lever is active.

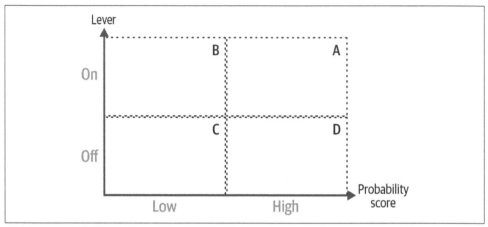

Figure 4-2. 2×2 test of a model and a lever

Note how the 2×2 design works here: you treat those users in groups A and B in the diagram, and the control group is composed of groups C and D. Variation in both dimensions allows you to do some testing of the lever and the model.

To get a real sense of the benefits of the design, imagine that you want to do a cross-selling campaign. For this, you trained an ML classification model that predicts who will accept an offer or not. If the model is predictive, high probability scores should have high true positive rates.

You want to test it using a new communication campaign that makes salient the benefits of buying the new product ("customers who use the new heart rate monitor feature on their smartwatch increase their running performance by 15%"). Let's also assume that the baseline campaign solely provided information on the new feature ("our new smartwatch includes state-of-the art monitoring for runners"). The metric for success is conversion rate (CR), measured as Purchases/Users in the campaign.

The hypotheses to be tested are as follows:

Monotonicity
    Higher probability scores have higher conversion rates: $CR(A) > CR(B)$ and $CR(D) > CR(C)$

Effectiveness

The new communication lever is more effective than the baseline: $CR(B) = CR(C)$ and $CR(A) > CR(D)$

The reason I expect $CR(D) > CR(C)$ is that some users make a purchase organically, without the need of having a communication displayed. If the model is predictive (in a true positive sense), the conversion rates should also be increasing with the score.

Similarly, I expect $CR(B) = CR(C)$ because I'm targeting users with a low probability of making a purchase, according to the model. It's true that great communication campaigns might convert some of these low-intentionality users, but I see no reason to expect the impact of the communication lever to be statistically significant.

To set up the experiment you must bring in statistical size and power considerations, where sample size and minimum detectable effects are critical. Usually, you don't have large enough samples, so an option is to just settle for having a good design for the lever (as in a classic A/B test framework), and a suboptimal design for your model. In this case, you may only have casual evidence of the model's performance. I've found this to be enough in most cases, but if you can go all in and have a good design for both factors, please do so. After the experiment is run, you can then test these hypotheses and get some evidence of model performance in real-life settings *and* the impact of a lever.

Example: Understanding User Behavior

I started discussing 2×2 *statistical* designs because thanks to the power of randomization, you control for other factors that may impact the metric of interest. Other use cases for the 2×2 framework generally lack this very nice property. Nonetheless, it may still be useful, as I hope this example shows.

Not so long ago I decided to set up a 2×2 framework to understand product-market fit for a specific product. For this, I took two factors that were critical for fit, and focused on quadrant A to single out those users that were doing great on both. I then built an ML classification model where users in group A were labeled with a one, and everyone else was labeled with a zero. The objective was to *understand* who these users were. In Chapter 13 I'll show how this can be done in practice, without the 2×2 framework.

In that particular use case, I used customer engagement and unit price. Group A consists of users who are highly engaged and are willing to pay high tickets. Engagement is usually a good proxy for product-market fit, so combining it with a proxy for revenue gave me what may be called *profitable fit*.

Let me give another example that applies the same logic. Recall that *customer lifetime value* (LTV) is the present value of a users' lifetime relationship with the company:

$$\text{LTV} = \sum_t \frac{r_t \times s_t}{(1+d)^t}$$

Here, r_t is the revenue at time t, s_t is the probability of surviving from *t–1* to *t*, and *d* is a discount rate. Sometimes, instead of revenue you can use a profit metric that also takes into account some form of costs, but in many companies, especially start-ups, it's common to use a revenue metric to compute the ratio of LTV to customer acquisition costs (CAC).[1]

As you can see, LTV can be expressed as the (discounted) inner product of revenue and survival probability streams. Suppose you want to understand what type of users have a high LTV. Who are they? What makes them so special? And most importantly, are there levers to move some users to the top LTV bucket?

Figure 4-3 shows the already familiar setting. On the horizontal axis I have a proxy for survival probability, and revenue is on the vertical axis. Since LTV is the inner product of streams at different time periods, you need to find ways to make both of these one-dimensional. There are several ways to do so, none without their own problems.

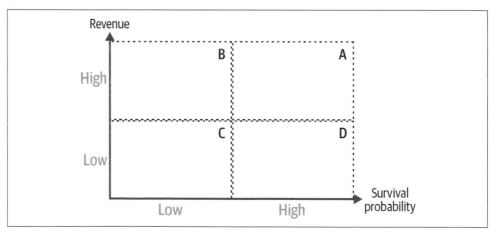

Figure 4-3. LTV in a 2×2 framework

---

1 Some companies also report the "undiscounted" LTV, so this expression simplifies to the summands on the numerator.

Forgetting about those details for now, you can proceed as I did in the previous example:

1. Label users in group A with ones and everyone else with zeros, and train a classification model that predicts being a quadrant A user.

2. Open the black box and try to learn something about users that have a high probability of being in quadrant A (using the methods presented in Chapter 13).

3. Scoring the complete user base, and using some threshold score, you can calculate the opportunity size for the product.

There are at least two methods to go from streams across time to two dimensions:

Aggregate.
> The simplest way is to use an aggregate statistic like average or median survival rates and revenue. Note that aggregation with sums might put at a disadvantage younger cohorts for revenue (for example, a user transacting for 20 months can generate 20× more revenue than a new user).

Choose an arbitrary period.
> If you've found in the past that the first six months are critical for survival (or revenue), you can just set this and use the corresponding values at that point in time.

Example: Credit Origination and Acceptance

A somewhat different example is the case of correlated outcomes. Take the case of a credit product (such as a credit card). These products are somewhat problematic because of *adverse selection* (riskier users are *more likely* to accept an expensive loan offer).

Figure 4-4 shows a somewhat typical scenario. Adverse selection creates the positive correlation, so users who are more likely to accept a loan offer are also more likely to default (A).

The 2×2 design simplifies the decision-making process: which customers should you target?

Offers in quadrant B.
> These customers are more likely to accept *and* to repay the loan. This is the safest move.

Adjust the thresholds to get more volume.
> You can also move the threshold definitions for low or high risk of default. This may help you find more volume if scale is of utmost importance. Credit

originators commonly do this type of calibration given their risk appetite. The 2×2 design lets you focus on one lever (risk threshold).

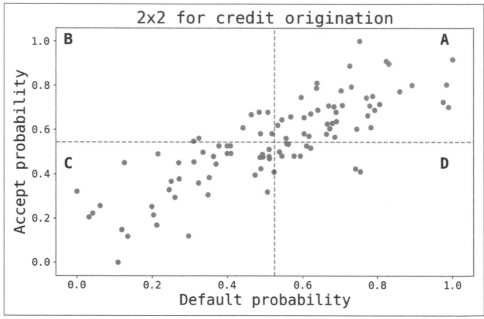

Figure 4-4. 2×2 loan origination example

Example: Prioritizing Your Workflow

A final example that is constantly used by consultants should help you prioritize projects. Here, the two dimensions used are value (of the project to the company) and how much effort is needed to complete it.

The idea is that you should rank competing projects along these two dimensions. In Figure 4-5, you can see that projects *x* and *y* are almost as good in terms of value, but *x* is to be preferred since it takes considerably less effort to complete. Similarly, ranking activities *y* and *z* is relatively easy since both require comparable efforts, but the former creates substantially more value. In general, the top left quadrant is where you want most of your projects to live.

As rich as this 2×2 view may be, it has its limitations. For instance, how do you compare projects *x* and *z*? In Chapter 5, I present an alternative that can be used more generally to compare and rank any set of projects.

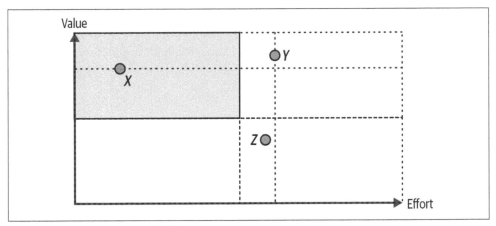

Figure 4-5. Prioritization of effort

Key Takeaways

These are the key takeaways from this chapter:

Case for simplification
The amount of data at your disposal notwithstanding, simplifying the world is necessary if the objective is to improve your understanding of a complex world and business. Moreover, it helps in communicating technical results to the stakeholders and allows you to focus on what appears to be of first-order importance.

2×2 diagrams
These tools simplify high-dimensional spaces into two-dimensional plots that allow you to focus on specific features or factors that are most relevant to the problem at hand.

Use Case 1: Testing a model and a lever
A common use case is 2×2 statistical designs. One such example is when you want to simultaneously test the effectiveness of a lever and the predictive performance of an ML model. You get crisp hypotheses that can go through the formal statistical testing process. Randomization guarantees that everything else remains constant, on average.

Use Case 2: Understanding your customers
By singling out two specific features, you can use the framework as a starting point for more sophisticated approaches. This chapter described how this framework can be used to understand which users have a high LTV.

Use Case 3: Correlated features

When there are correlated features, the 2×2 framework allows you to simplify the decision-making process. The example I used was loan origination, where offer acceptance depends on the probability of default because of adverse selection.

Further Reading

In my book *Analytical Skills for AI and Data Science*, I argue that learning to simplify is an essential skill for data scientists. The discussion is way more general than this chapter, and I did not cover 2×2 designs. I also discuss LTV and the design of A/B tests.

The Laws of Simplicity by John Maeda (MIT Press) takes a designer's point of view on how to achieve simplicity. While it may sound unrelated, I've found that somewhat orthogonal points of views have always deepened my understanding of a problem.

2×2 statistical designs can be found in most statistical textbooks where ANOVA is covered. *Statistical Methods in Online A/B Testing: Statistics for Data-Driven Business Decisions and Risk Management in E-Commerce* by Georgi Zdravkov Georgiev (independently published) has a good discussion of testing with multiple variants and other related topics.

The type of adverse selection used in the loan origination example is covered in any microeconomic textbook discussing information asymmetries. If you do not have an economics background, the technical nitty-gritty might be an overreach. In my opinion, the important part to remember is that users self-select using information about themselves that is not known by the decision-maker, and this creates a lot of problems.

## CHAPTER 5

Building Business Cases

Learning to write a business case for a model or experiment is a critical skill that data scientists ought to develop. Not only can it help you quickly learn whether a new project is worth your time and effort, but it can also help you gain stakeholder buy-in. Moreover, it is consistent with the type of extreme ownership that will make you shine.

Business cases can be as complex as you want, but many times you can come up with good-enough estimates. In this chapter, I will go through the fundamentals of business case creation.

Some Principles to Construct Business Cases

While every business case is different, most can be built using the same underlying principles: you compare making a decision or not, calculate costs and benefits of all options, consider only incremental changes, and many times you can account only for unit economics.

Decisions
> Business cases are most commonly built to evaluate a new decision that is under consideration, be it a new campaign, a change in a lever, or any other decision.

Costs, benefits, and breakeven
> Most interesting decisions have trade-offs. A critical starting point is to enumerate the main costs and benefits derived from the decision. The business case will be built around *net benefits* calculated as the monetary difference between benefits and costs. *Breakeven* is synonymous with having zero net benefits and serves as the limit case, or worst-case scenario, for your decision.

Incrementality

A good business case should only take into account those costs and benefits that arise *from* the decision. For example, your salary can be viewed as a cost if you're running an experiment, but this is not *incremental* since the company would also have to pay you if you were doing something else. Only incremental costs and benefits should be included.

Unit economics

Most of the time it only matters what happens to your *average* customer, so you can just focus on the incremental costs and benefits for this isolated unit. The business case depends on the sign of the net benefits you calculate for this unit; usually, scaling to the whole customer base affects both costs and benefits in the same proportion, leaving the sign of aggregate net benefits unaffected.

Example: Proactive Retention Strategy

Let's evaluate whether the company should launch a proactive retention strategy. On the costs side, you need to give a customer an incentive to stay. There are many ways to do this, but most can be easily translated to a monetary figure c. On the benefits side, a customer that stays for one extra month generates average revenue per user r that was going to be lost.

Suppose you target a customer base of size B. Of these, A accept the incentive. Also, out of those targeted, only TP were really going to churn (true positives). The break-even condition is obtained by equalizing costs and benefits:

$$B \times \frac{A}{B} \times c = B \times \frac{TP}{B} \times r$$

You can see one of the techniques presented in Chapter 2 at play. Notice how, in this case, you can focus only on the average unit:

$$\frac{A}{B} \times c = \frac{TP}{B} \times r$$

It makes sense to run the campaign when the net benefit is nonnegative:

$$\frac{TP}{B} \times r - \frac{A}{B} \times c \geq 0$$

The first fraction is just the true positive rate in the campaign base or sample; the second fraction is the acceptance rate. Alternatively, and conveniently, you can also view these as sample estimates for the expected benefit and cost, so that your decision problem maps neatly to one under uncertainty: before the campaign you don't know who will accept the incentive or who will actually churn in its absence.

You can now plug in some numbers to simulate the business case under different scenarios. Moreover, you can also analyze the levers at your disposal. Here there are three levers for the business case to work:

Improve the true positive rate.
You can help the business case by making more accurate predictions with your machine learning (ML) model, in a true positive sense.

Keep costs under control.
You can lower the value of the incentive (c). Sometimes it's safe to assume that the acceptance rate increases with it, so both terms go in the same direction.

Target only customers with high ARPU.
It makes intuitive sense that incentives should be prioritized to high-value customers. In the inequality, this corresponds to a higher r.

Note how incrementality kicks in: on the benefits side, you should only include the *saved* ARPU from those customers who were really going to churn (true positives). Those who were going to stay, independently of the incentive, increase the cost if they accept but provide no incremental benefits.

What about *false negatives*? Remember these are customers that are not targeted and churn. You can include the lost revenue as a cost so that you can trade off precision and recall in your ML implementation.

Fraud Prevention

Banks frequently establish transaction limits for fraud prevention purposes (and for anti-money laundering). Let's build a business case for the decision to block a transaction whenever it exceeds the limit.

Intuitively, there are two costs: fraud cost (c_f) and the lost or forgone revenue if a customer churns (c_{ch}). For simplicity I will assume that a customer with a blocked transaction churns with certainty, but this assumption is easy to relax in applications. On the revenue side, if a transaction is allowed to go through, the company gets the ticket amount (t).

Once a transaction comes in, you can either accept or block it. Independently of the action, it can be legitimate or not. Table 5-1 shows costs and benefits for all four combinations of actions and outcomes.

Table 5-1. Costs and benefits for fraud prevention

Action	Outcome	Benefits	Costs
Accept	Fraud	t	c_f
Accept	Legit	t	0
Block	Fraud	0	0
Block	Legit	0	c_{ch}

Denote by p the probability that a given transaction is fraudulent. Computing the expected net benefits from each possible action, you get:

$$E(\text{net benefits}|\text{accept}) = p\left(t - c_f\right) + (1 - p)t = t - pc_f$$

$$E(\text{net benefits}|\text{block}) = -(1 - p)c_{ch}$$

Blocking a transaction with ticket t is optimal whenever net benefits from blocking exceed those from accepting the transaction:

$$E(\text{net benefits}|\text{block}) - E(\text{net benefits}|\text{accept}) = pc_f - \left(t + (1 - p)c_{ch}\right) \geq 0$$

This last inequality is at the heart of the business case. On the benefits side, if the transaction is fraudulent, by blocking you save the cost from fraud (c_f). On the cost side, by blocking a transaction you effectively neglect the revenue t and incur the potential cost of churn (c_{ch}) if the transaction is not fraudulent.

As before, let's turn the attention to the levers. Other than blocking or accepting, you can always choose the limit (L) such that higher tickets will get blocked and anything else will be accepted. But where is the limit in this inequality?

The probability of being fraudulent is usually a function of this limit: $p(t|L)$. In many applications it's common that this function is increasing in the limit; this arises whenever fraudsters are looking for short-term and quick, relatively large rewards. By setting a sufficiently large limit, you can then focus on high probability transactions. The cost of fraud is usually the ticket itself, so there's also this direct effect on the benefits. There is a trade-off, however: if a transaction is not fraudulent, you risk the churn of high-value customers.

Purchasing External Datasets

This logic applies for any decision you want to analyze. Without going into details, I'll briefly discuss the case to purchase an external dataset, a decision that most data science teams evaluate at some point.

The cost is whatever your data provider decides to charge for it. The benefit is the incremental revenue your company can create with the data. In some cases this is straightforward since the data itself improves the decision-making process. I'm thinking of use cases like KYC (know your customer) or identity management. In cases like these you can map the data to revenue almost one-to-one.

In most other cases that are interesting from a data science perspective, the incremental revenue depends on critical assumptions. For example, if you already have an ML model in production used in the decision-making process, you can quantify the minimum incremental performance that makes the business case positive, given this cost. Alternatively, you can try to negotiate better terms, given this incremental performance.

The idea can be summarized by something like this:

$$\text{KPI(augmented dataset)} - \text{KPI(original dataset)} \geq c$$

The KPI is a *function* of your ML model performance metric. I emphasize the function part because you need to be able to convert the performance metric into a monetary value, like revenue, to make it comparable with the costs. Note that by using the original dataset as a benchmark, you consider only incremental effects.

Working on a Data Science Project

As suggested in Chapter 1, data scientists should engage in projects that are incremental for the company. Suppose you have two alternative projects, A and B. Which should you start with? Using the same logic, you should choose project A if:

$$\text{revenue}(A) - \text{cost}(A) \geq \text{revenue}(B) - \text{cost}(B)$$

To make a decision you need to plug in some numbers, for which the calculation is a project in and of itself. What matters here is the intuition you get from the inequality: prioritize those projects for which there's substantial incremental net revenue, given *your* implementation costs.

In Chapter 4, I showed how a simple 2×2 framework can help you prioritize your workflow by ranking each project on the value and effort axes. As useful as it is, with this graphical device you may end up having trouble ranking projects that dominate in one dimension and get dominated in the other dimension (for example, projects x and z in Figure 4-5). The previous inequality solves this problem by using a common scale (money) to value effort (cost) and revenue.

Key Takeaways

These are the key takeaways from this chapter:

Relevance
> Learning to write business cases is important for stakeholder management and extreme ownership purposes, as well as for allocating data science resources across alternative projects.

Principles of business case writing
> Typically, you need to understand cost and benefits, as well as breakeven. Focus only on incremental changes. Many times, you only need to care about unit economics affecting your average customer.

Further Reading

In my book *Analytical Skills for AI and Data Science*, I describe techniques that will help you simplify your business case to focus only on first-order effects. It will also help you understand decision making under uncertainty.

This cost-benefit analysis is standard in economic analysis. What I've labeled here as *incrementality* is commonly known as *marginal analysis*. Three books that I'd recommend to noneconomists are: *The Armchair Economist: Economics and Everyday Life* by Steven E. Landsburg (Free Press), *Doughnut Economics: Seven Ways to Think Like a 21st-Century Economist* by Kate Raworth (Chelsea Green Publishing), and *Naked Economics: Undressing the Dismal Science* by Charles Wheelan (W. W. Norton).

## CHAPTER 6

What's in a Lift?

There are very simple techniques that help you accomplish many different tasks. Lifts are one of those tools. Unfortunately, many data scientists don't understand lifts or haven't seen their usefulness. This short chapter will help you master them.

Lifts Defined

Generally speaking, a *lift* is the ratio of an aggregate metric for one group to another. The most common aggregation method is taking averages, as these are the natural sample estimates for expected values. You'll see some examples in this chapter.

$$\text{Lift}(\text{metric}, A, B) = \frac{\text{Metric aggregate for group } A}{\text{Metric aggregate for group } B}$$

In the more classical data mining literature, the *aggregate* is a frequency or probability, and group *A* is a subset of group *B*, which is usually the population under study. The objective here is to measure the performance of a selection algorithm (for example, clustering or a classifier) relative to the population average.

Consider the lift of having women as CEOs in the US. Under a random selection baseline, there should be roughly 50% female CEOs. One study (*https://oreil.ly/27yD1*) estimates this number at 32%. The lift of the current job market selection mechanism is *0.32/0.5 = 0.64*, so women are *underrepresented* relative to the baseline population frequency.

As the name suggests, the lift measures how much the aggregate in one group increases or decreases relative to the baseline. A ratio larger or smaller than one is known as *uplift* or *downlift*, respectively. If there's no lift, the ratio is one.

Example: Classifier Model

Suppose you train a classifier to predict customer churn. You have a dataset where users who churned are labeled with a one, and those who are still active are labeled with a zero. The baseline churn rate is obtained by taking the sample average of the outcome.

One common performance metric to track is the true positive rate by score decile in the test sample, which translates to churn rate by decile in this example. To compute it, you just need to sort the users by score and divide the test sample into 10 equally sized buckets or deciles. For each bucket, compute the churn rate.

This metric is useful because it informs you about at least three important aspects:

Lifts
> Dividing the churn rate per decile by the churn rate in the test sample, you compute the corresponding lifts. This is an estimate of how well the model is identifying churners in each decile relative to the company's churn rate.

Monotonicity
> Is the score informative? If the probability score is informative, in a true positive sense, higher scores should have higher churn rates.

Top decile performance
> In many applications, you just target users in the highest decile. In this example, you may only want to give a retention incentive to those who are likeliest to churn. The true positive rate for that decile is your first estimate of what can be expected in the retention campaign.

Figure 6-1 shows *true positive rates* (TPRs) and lifts for a simulated example. The classifier identifies churners at 2.7× the average rate in the top decile. This is a good finding if you want to convince your stakeholder to use the output from your model. You can also benchmark this lift against the one obtained through their current selection mechanism.

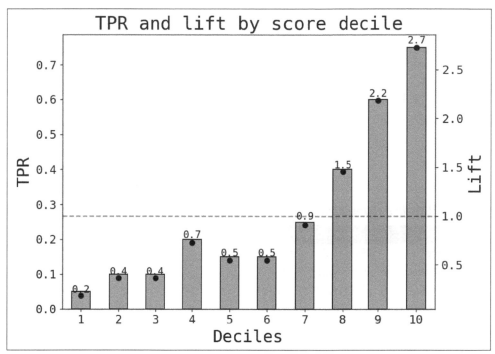

Figure 6-1. TPR and lifts for the churn model example

Self-Selection and Survivorship Biases

Self-selection arises when an individual *chooses* to enroll in a group. Examples can be groups with formal enrollment (such as a political party or a team) or informal enrollment (for example, purchasers of your product, users of a feature, and the like). The important thing is that there is some intrinsic characteristic that drives the individual to become a member.

Survivorship bias is a type of self-selection in reverse: some users end up in your sample ("survive") because of some characteristic they have. The classical example (*https://oreil.ly/0Y9oW*) is the case of World War II fighter planes analyzed by statistician Abraham Wald. The lesson is that you may end up with incorrect conclusions because of the biased nature of your sampling process.

Chapter 15 discusses the relevance of self-selection bias for data scientists; for now, it suffices to show how lifts can help you quickly identify the presence of this bias.

Table 6-1 shows the typical way this is presented: the rows include some features or characteristics you believe are important to understand the selection problem; columns highlight group membership as well as lift. Here I only include four variables for a customer:

- Monthly spend on the company's products
- Satisfaction score
- Monthly income
- Tenure

Table 6-1. Lifts in a churn example

	Active	Churned	Lift
Monthly spend	29.9	32.7	1.1
Customer satisfaction score	10.00	10.08	1.01
Income (K)	46.52	54.80	1.18
Tenure (months)	9.84	8.14	0.83

Generally, the more features you have, the better the understanding of your selection mechanism that you can get with lifts. For instance, why not include geographical or industry segments or the number of products the customer has already purchased from the company?

Each cell in the table shows the average of the corresponding feature for active and churned users, as well as the lift. For instance, average spend for active and churned users is $29.9 and $32.7, respectively. Looking at the lift column, it's easy to recognize a pattern: churners have a higher income (lift is 1.18, or an 18% increase), spend more (1.1), and have been a customer for a shorter time (0.83). Customer satisfaction scores are not important (negligible uplift). One possible story for these findings is that relatively better-off customers have higher expectations for the products; this seems to be a product for a lower socioeconomic status segment.

In any case, you get the idea: a quick and dirty approach to understanding a selection mechanism is to construct lift tables. If features are selected correctly, you can immediately get a sense of what's going on with the underlying group.

Other Use Cases for Lifts

The technique is very simple to use: identify a metric and groups, and compute the ratio. The selection mechanism can be anything you find relevant.

For instance, you can use the 2×2 diagrams presented in Chapter 4 and focus on one of the quadrants. Lifts are very simple to use and might help you understand what drives the users in that group.

Another common use case is when analyzing self-selection in marketing campaigns. In the absence of selection bias, you can measure the impact of a campaign using a control group. Lifts will very quickly let you know if you can proceed that way or not.

Similarly, many surveys end up with biased results because of differences in response rates across groups. In the past, I've automated representativeness checks for customer satisfaction surveys using lifts.

Key Takeaways

These are the key takeaways from this chapter:

Defining lifts
> A lift is a ratio of an aggregate metric for one group to another. Averages are the most common aggregation method.

Lifts in machine learning
> You can compute lifts with classifier models by showing the predictive performance of the model relative to the overall sample. I presented a churn prediction example and computed lifts for true positive rates across score deciles.

Self-selection
> More generally, lifts can be applied to understand the extent of self-selection or survivorship bias in your sample. By computing the lift of a metric in those users who self-selected themselves into a group, you can easily understand the selection drivers.

Further Reading

Lifts are covered in many classical data mining books; for instance, see *Data Mining: Practical Machine Learning Tools and Techniques* by Ian Witten et al. (Morgan Kaufmann).

More references can be found in academic articles and the blogosphere. Examples are "Lift Analysis—A Data Scientist's Secret Weapon" by Andy Goldschmidt on KDnuggets (*https://oreil.ly/KfBaL*) and "ROC Curve, Lift Chart and Calibration Plot" by Miha Vuk and Tomaz Curk (*Metodoloski Zvezki* 3 no. 1, 2006: 89–108).

## CHAPTER 7

Narratives

You've spent weeks working on your project and are now ready to present the results. It feels like you're almost done, and *just* have to deliver the output.

Many data scientists think this way, and put little to no effort into building compelling narratives. As described in Chapter 1, to have end-to-end ownership, it is critical to persuade your stakeholders to take action with your results. This type of extreme ownership is critical to create value; hence, you must master the art of storytelling.

There are plenty of resources out there to learn about storytelling (I'll suggest some at the end of this chapter). This chapter builds on that body of knowledge, but I will deviate slightly to highlight some skills that are specific to data science.

What's in a Narrative: Telling a Story with Your Data

Using a standard dictionary definition, a *narrative* is just a sequence of connected events. These connections make a story. I will enrich this definition by saying that it should also accomplish an objective.

What is the objective that you want to achieve? In general narratives, it could be to persuade or engage. These apply also to data science (DS), of course, but most importantly, you want to create value, and for that you need to *drive actions*. A successful story should help you accomplish this objective.

Let's reverse engineer the problem and identify conditions that help us achieve this:

- Clear and to the point
- Credible
- Memorable
- Actionable

Clear and to the Point

Clarity is a relative concept that varies with the context and depends very much on your audience. Technical details of your machine learning (ML) implementation can be very clear to your data science team, but they are usually cryptic to your business stakeholder. Identifying your audience is a first critical step in building a clear narrative. Choosing the right language and tone for the right audience is thus critical.

DS is inherently a technical subject. As such, data scientists are very often tempted to include fancy technical jargon in their presentations (even better if there are some accompanying equations). But delivering a story is not about you. It's always good advice to put all the technical material in a technical appendix section, if you want to include one.

A common mistake is to think that technical language will buy you credibility (more on this later). Sometimes this comes in the form of trying to prove that the data science toolkit is necessary for the organization. My advice is to balance this desire against the benefits of having an effective communication that accomplishes your objective. Creating powerful narratives is about the latter.

In a regular DS development process, it's quite normal to run many tests and create multiple visualizations. When trying to make a case for the amount of work they have put into it, some people are tempted to include everything they can in the presentation, thereby distracting and overwhelming the audience. Focus only on the key messages, and include results that reinforce them. Everything else should be dropped. If something is not directly helping your delivery, but might still be useful, put it in the appendix. But try not to clutter the appendix; this section also serves a specific purpose in your presentation (if not, drop it).

Achieving the right amount of simplicity takes a lot of practice and effort; it's a skill in its own right. A good tip is to start writing what you think are the key messages, and then start dropping everything else from the presentation. Iterate until convergence: stop when you've dropped so much that the message is not clear anymore.

This advice also applies to sentences and paragraphs. Use short sentences, with fewer than 10 words, if possible. Long sentences and paragraphs are visually exhausting, so you can assume they won't be read. Once I have a first draft, I go through each sentence and paragraph and make them as short and clear-cut as possible.

Clarity should be channel independent. Many times you prepare for a live presentation and fail to recognize that part of the audience—possibly C-level—will *read* it at some other point in time. You must therefore make it self-explanatory.

This applies not only to text but also to data visualizations. Be sure you label all relevant axes and write meaningful titles. If there's something you want to highlight in a specific figure, you may include visual aids—such as highlighting, text, or boxes—to help direct your audience's attention.

Data visualizations are an intrinsic part of the delivery of data science narratives. These principles apply to any figures you prepare. I will cover some good practices for data visualizations in Chapter 8.

Tips to Achieve Clarity

Here are some key tips to achieve clarity:

Audience
Start by identifying your audience and guaranteeing that the language and tone are consistent.

Technical jargon
Control your temptation to include technical jargon. Always put the technical material in the appendix.

Focus on the key messages
The key messages should be clear from the outset, and the narrative should be built around them. Anything else should be dropped.

Delete distractions
Write your first draft, then start dropping anything that's not necessary. This also applies to sentences and paragraphs.

Self-explanatory
Narratives should always be self-explanatory, independent of how you deliver them.

Datavis
Apply all of the above tips to your data visualizations.

Credible

In business settings, compelling narratives must be credible. Unfortunately, this is a very subtle property: it takes time to gain, but is terribly easy to lose. In data presentations there are three dimensions that you should care about:

- Data credibility
- Technical credibility
- Business credibility

Data quality is at the core of the first dimension, and you should make it a practice to include checks at the source as well as during the development cycle. Data scientists write a lot of code, an error-prone activity. The worst happens when your code actually runs, but the results may just not be right (logic errors). In my experience, this happens *a lot*. The best programmers have made testing an intrinsic part of their day-to-day workflow.

Moreover, data can't be disentangled from its context, so your results must make sense from a business perspective. I've seen many data scientists lose their credibility because they didn't check that their results made sense. At a minimum, you should know the order of magnitude for the key metrics you're working with. At best, you should know the metrics by heart. Don't forget to always challenge your results before presenting them.

Technical credibility is usually granted by stakeholders. But great powers come with great responsibility. Data scientists need to learn to use the right tool for each problem, and master the techniques. It's always a good practice to have an internal seminar series where you can be challenged by your peers.

As described in Chapter 1, it's crucial to show business expertise with your audience. Your technical delivery might be impeccable, but if things don't make sense from a business standpoint, you'll lose credibility with your stakeholders. I've seen data scientists start with faulty assumptions about how the product or business works. Another common mistake is to make incredible assumptions about the customers' behavior; always ask yourself if you would do this or that if *you were the customer*.

Tips to Achieve Credibility

Here are some tips to achieve credibility:

Data credibility
> Check your results to ensure that they make sense from a business perspective.

Technical credibility
> When possible, make it a practice to present your technical results to knowledgeable peers.

Business credibility
> Aim for being as knowledgeable about the business as your stakeholder.

Memorable

In ordinary narratives this is usually achieved by injecting some form of struggle or suspense, effectively making the sequence less linear. I learned the hard way that these are generally not good strategies for DS narratives.

In DS, memorability is most often created by an *Aha! moment*. These usually arise when you show an unexpected result that can create value. This last part is important: in business settings, intellectual curiosities are only short-term memorable. The best Aha! moments are those that drive actions.

Many authors suggest creating unforgettable narratives by using the right combination of data and emotions, and there is indeed evidence that the human brain recalls what tickles your heart better than plain scientific evidence. I agree with this general point, but in my opinion it's not the best practice to aim at TED-type narratives and delivery. You should rather keep it simple and find actionable insights that are almost always memorable by themselves.

Aha! moments, in the form of actionable and somewhat unexpected insights, are the best way to achieve memorability.

Actionable

If you've gone through the previous chapters, it shouldn't come as a surprise that I believe this ought to be your North Star and the main character in your story.

Before starting your narrative, ensure that your project has actionable insights. If not, go back to the drawing board. I've seen presentations where something interesting is shown, but the audience is left thinking *So what?*

Identify levers that arise from your analysis. A presentation without actionable insights cannot be effective at creating value.

Building a Narrative

The previous section presented some properties that are necessary to create successful narratives, as well as some tips that will help you ensure that these are satisfied in practice. I'll now cover the process of creating narratives.

Science as Storytelling

Many data scientists think of storytelling as something independent, even orthogonal, to their technical expertise, only exercised at the delivery stage. My claim is that data scientists should become *scientists* and make it an inherent part of their end-to-end workflow.

With this in mind, let me suggest two alternative processes to building a narrative:

- First, do the technical work, and *then* create the narrative.
- Start with an initial narrative, develop it, iterate, and when you are ready, sharpen the storytelling for delivery.

The first process is the most common in data science: practitioners first do the hard technical work and then *fit a narrative* to their results. Usually, this ends up being a collage of findings that may be interesting or relevant but lack a story.

In contrast, the second process makes storytelling an integral part of the data science workflow (Figure 7-1). In this setting you start with a business question, do the hard work of understanding the problem and come up with some stories or hypotheses, test them with data, and after some iterations, you're finally ready to deliver the results. This process may even force you to go back and redefine the business question.

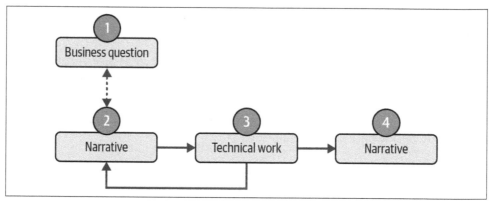

Figure 7-1. Iterative narratives

Storytelling is present at the beginning (stage 2), in the middle as you iterate and refine your hypotheses (stage 3), and at the end (stage 4). These narratives are not quite the same, but are certainly connected.

You, and your need for understanding, are the audience for the initial narrative; your business stakeholders are the audience for the final one. As mentioned, the language and tone are different. But the important thing is that the *key messages* at the end are direct descendants of those at the beginning, purified by the iterative process in the middle.

You should come into a project with a first candidate set of key messages that you think will be delivered at the end. These may not be entirely right, but more often than not—if you have enough business expertise—they aren't too far off from the final ones. With this approach, the process of creating a narrative for your presentation starts *before* you even get to the data. It will guide you *during* this stage and will help you iterate. Thanks to this intermediate stage, you're able to catch mistakes and find errors (in data quality, logic, coding, or even your understanding of the business); this is usually the place where Aha! moments arise. The delivery-stage narrative switches audiences and communicates the *final* and refined messages.

What, So What, and Now What?

Once you've reached the delivery stage, you need to put some structure onto your narrative. Some people like to follow the standard approach to storytelling—also known as the *narrative arc*—that has three acts: setup, struggle, and resolution.

While this may work for some, I prefer a different sequence that reinforces your key objective of driving actions: what, so what, and now what? Not surprisingly, this corresponds closely to the process described in Chapter 1.

What?

This section is about describing the business problem and its importance to the company *at this point in time*. It should also include some quantitative information on the context, like the recent evolution of the main KPIs and opportunity sizing.

Imagine that you are trying to quantify the impact of giving price discounts. In this section you can provide some context, like the recent frequency of price changes, the range or distribution, and some high-level impact on sales, retention, or revenue. If there's some evidence, even if casual, you can also highlight the strategic importance in the current competitive landscape.

So what?

The critical thing about this section is to focus on actionability. The main results are going to be here, including those that generate Aha! moments.

Generally speaking, Aha! moments come in two flavors:

- Unexpected results
- Somewhat expected results (directionally speaking) but with an unexpected twist that can come from quantification or actionability

I prefer the second type because you should have a bias for action. If you have unexpected results *and* an action plan, you nailed it.

Going back to the pricing example, dropping prices usually boosts sales. This is the expected behavior, so showing a negative correlation won't create an Aha! moment, and the audience may end up with the feeling that you are reinventing the wheel.

But if you say that users are relatively price insensitive for prices $5.30 or higher, but that below that price, each additional $1.00 discount boosts sales by *1,000* units, you have captured their attention. The message is similar, but there's a surprise component that comes from quantifying things. Moreover, this is a call to action that needs to become the centerpiece of the last section.

Now what?

This section is about next steps. What do you need from the rest of the company to capture this value? Who needs to be involved? Here, I like to actually suggest specific next steps. I've seen data scientists be shy about this because they aren't usually the actual decision-makers.

With the pricing example, most likely you rely on the marketing team to design and communicate the actual discount strategy. Finance may also need to approve the plan. Any other teams affected should also be included.

The Last Mile

In her book *The Hard Truth About Soft Skills* (Harper Business), Peggy Klaus suggests that long-term success is 75% about soft skills and the rest is technical knowledge. I'm not sure if that's right, but directionally speaking I cannot agree more: data scientists invest substantial time and effort in achieving technical excellence, but their careers depend more on those soft skills that have been neglected.

In the last mile, it's now time to switch from your scientist to your salesperson persona. I've learned from personal experience that many great projects fail because of lack of preparation at this stage.

Writing TL;DRs

TL;DRs (too long; didn't read) are a great tool to check whether your narrative is sharp and concise enough. They have become a standard in tech companies, and I've made it a practice to always start with one.

Many executives won't spend much time on your work *unless* they see something that catches their attention. Great TL;DRs are written to achieve that.

Tips to Write Memorable TL;DRs

Some people like to write a first draft of the TL;DR *before* writing down the actual document. This is a great way to ensure that the TL;DR is consistent with your narrative and guarantee that the contents are aligned with it. After finishing, they go back and iterate on the TL;DR.

The approach I prefer is to write down the narrative first (some people draw an actual sketch in pen and paper), work on the contents, and only then go back and write the TL;DR. To me, the TL;DR is the last thing you write, and I always sketch the narrative first.

The two approaches may sound similar, but the TL;DR is a really sharpened version of the narrative. The narrative is a high-level view of the story tying the sequence of events; the TL;DR is its laser-focused version.

I tend to structure the TL;DR in the same way as the narrative: What, So What, Now What. As before, the What section makes the case for your audience's attention, the So What summarizes the main findings and actionables, and the Now What are suggested next steps.

 A good tip is to think of your document as a news article and think about alternative headlines. Great headlines in data science must have the same properties I've been talking about: simple, credible, memorable, and actionable. Credibility restricts you from overselling.

Finally, everything in your TL;DR should have a slide that expands on it. If it was important enough to make it to the TL;DR, you better have some accompanying material.

Example: Writing a TL;DR for This Chapter

Figure 7-2 shows an archetypical TL;DR you might encounter. It's cluttered, and, evidently, I was trying to include every single detail of the work. It has very long sentences, and a small enough font to make it fit the page. It's certainly not readable. It's also memorable, but for the wrong reasons.

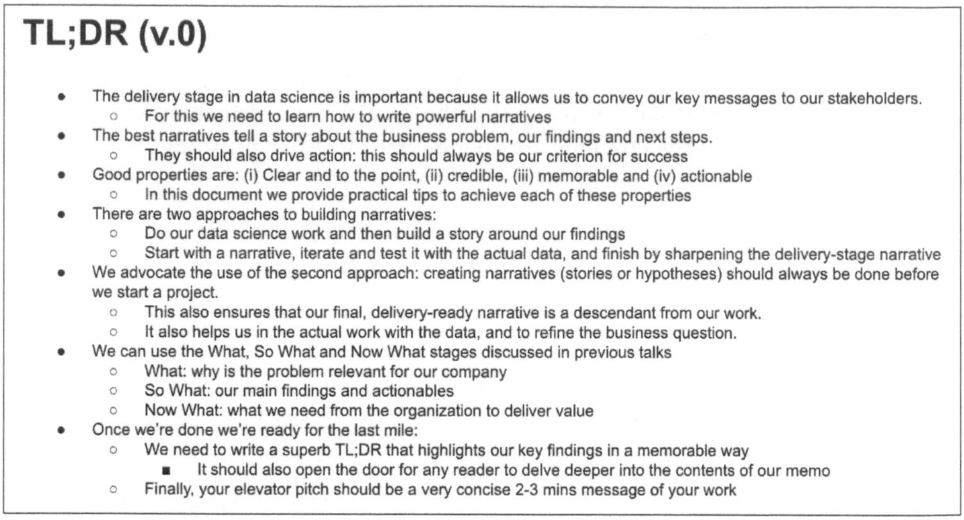

TL;DR (v.0)

- The delivery stage in data science is important because it allows us to convey our key messages to our stakeholders.
  - For this we need to learn how to write powerful narratives
- The best narratives tell a story about the business problem, our findings and next steps.
  - They should also drive action: this should always be our criterion for success
- Good properties are: (i) Clear and to the point, (ii) credible, (iii) memorable and (iv) actionable
  - In this document we provide practical tips to achieve each of these properties
- There are two approaches to building narratives:
  - Do our data science work and then build a story around our findings
  - Start with a narrative, iterate and test it with the actual data, and finish by sharpening the delivery-stage narrative
- We advocate the use of the second approach: creating narratives (stories or hypotheses) should always be done before we start a project.
  - This also ensures that our final, delivery-ready narrative is a descendant from our work.
  - It also helps us in the actual work with the data, and to refine the business question.
- We can use the What, So What and Now What stages discussed in previous talks
  - What: why is the problem relevant for our company
  - So What: our main findings and actionables
  - Now What: what we need from the organization to deliver value
- Once we're done we're ready for the last mile:
  - We need to write a superb TL;DR that highlights our key findings in a memorable way
    - It should also open the door for any reader to delve deeper into the contents of our memo
  - Finally, your elevator pitch should be a very concise 2-3 mins message of your work

Figure 7-2. TL;DR version 0

In Figure 7-3 I applied some of the tips given earlier to reduce the clutter: simplify and cut some of the sentences. Had I wanted to, I could have increased the font size. I could've done more, but realized that the best thing was to go back to the drawing board and start from scratch.

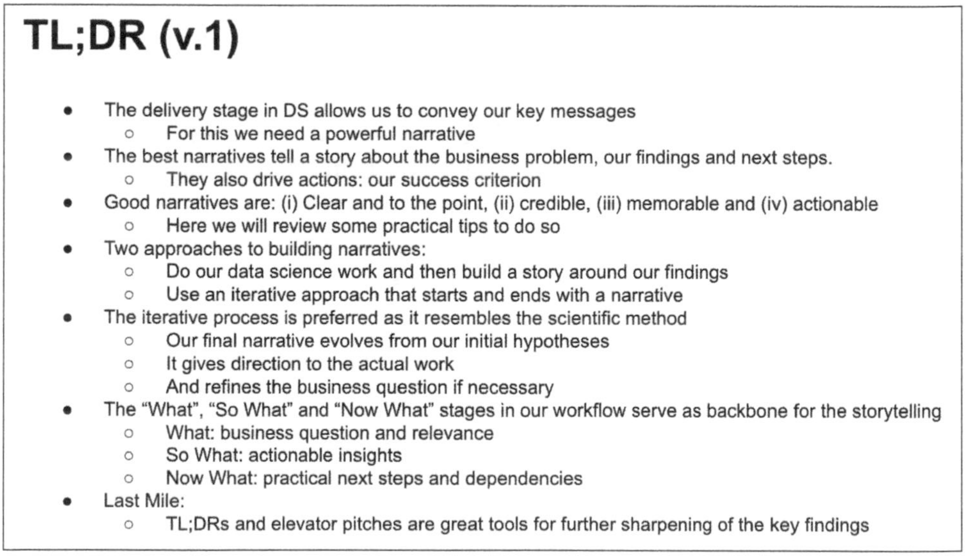

TL;DR (v.1)

- The delivery stage in DS allows us to convey our key messages
  - For this we need a powerful narrative
- The best narratives tell a story about the business problem, our findings and next steps.
  - They also drive actions: our success criterion
- Good narratives are: (i) Clear and to the point, (ii) credible, (iii) memorable and (iv) actionable
  - Here we will review some practical tips to do so
- Two approaches to building narratives:
  - Do our data science work and then build a story around our findings
  - Use an iterative approach that starts and ends with a narrative
- The iterative process is preferred as it resembles the scientific method
  - Our final narrative evolves from our initial hypotheses
  - It gives direction to the actual work
  - And refines the business question if necessary
- The "What", "So What" and "Now What" stages in our workflow serve as backbone for the storytelling
  - What: business question and relevance
  - So What: actionable insights
  - Now What: practical next steps and dependencies
- Last Mile:
  - TL;DRs and elevator pitches are great tools for further sharpening of the key findings

Figure 7-3. TL;DR version 1

Figure 7-4 shows the results of this last iteration. Starting from scratch allowed me to focus on the key messages. You can see that I followed the what, so what, now what pattern. In a real data science TL;DR, I would've highlighted some key results that are both quantified and actionable. The only call to action here is to practice.

TL;DR Creating narratives in data science (v.2)

- Powerful narratives serve three purposes
  - Structuring of a project
  - Focus on key results from the beginning
  - Drive action

- Narratives ought to be:
  - Simple
  - Credible
  - Memorable
  - Actionable

- In this talk we provide *practical* guidance to achieve these

- Data *science* as storytelling:
  - Start with a set of hypotheses
  - Test and refine them (data)
  - Delivery-stage narrative with key messages
  - Last mile

- Like any other skill, it's now time to put in practice

Figure 7-4. TL;DR version 2

It's also evident that I'm using a bullet point style. There are many detractors to this approach, but as everything it has its pros and cons. On the con side, it certainly restricts your creativity (imagine all you can do with a blank piece of paper). On the pro side, it forces you to write in a simple, clear, and orderly way. I can quickly see if my sentences are too long (sentences that take two lines are to be avoided, if possible).

As I said before, I don't think that TED-style presentations are a good fit in data science or business settings. Nonetheless, if you're skillful enough and it fits your company's culture, go right ahead. But bullet points tend to work well in business settings.

Delivering Powerful Elevator Pitches

Here's a trick that I learned some time ago when I was presenting to my manager: if someone starts a presentation and it's pretty obvious that it lacks a narrative, interrupt and ask that person to give you the elevator pitch. More often than not, there's no elevator pitch.

Elevator pitches are supposed to be the 10-to-20-second presentation that you would give to the CEO if you happen to meet on the elevator. You really want to sell your work! But there's a catch: you only have until you get to your floor. At that point, you've lost your chance to interact.

This has only happened to me once or twice, so I don't think of elevator pitches literally. Rather, I think of them as part of the narrative creation toolkit. Good narratives should be easy to summarize in an elevator pitch form. If you can't, most likely you have a problem with your story, and it's time to iterate.

 The next time you're working on your project, try your elevator pitch *before* and *after* you think you're done. That's your litmus test.

Presenting Your Narrative

Here are some good tips for the delivery stage:

Ensure that you have a well-defined narrative.
> If you followed the iterative approach, the narrative was always there, and you just need to discipline yourself. If you didn't, sketch the narrative *before* starting your deck or memo. When you're done, have someone go through your slides and ask for their version of the narrative. The narrative should be apparent to anyone if they only focus on the key messages per slide. If they can't identify it, you need to go back to the drawing board. There should also be clear and natural transitions between these messages.

Each slide should have a clear message.
> If a slide doesn't have a clear message consistent with your narrative, drop it.

Always practice giving the presentation.
> This is always true, but *especially* so if your audience includes top executives in the organization (and you should want this to be the case). A good practice is to record yourself: not only will this help you manage your time, but it will also help you identify any tics and mannerisms you may have.

Time management.
> Before presenting, you should already know how long it takes you *without* interruptions, so you better plan for the extra time. Also remember that you are the sole owner of your presentation, so you're entitled to (kindly) move on from questions that are taking you away from the key messages.

Quantify whenever you can, but don't overdo it.
> It goes without saying that DS is a quantitative field. Very often, however, I see data scientists describing their results in *qualitative* or directional terms. Instead of saying "Bad user experience increases churn," put some numbers into that

statement: "Each additional connection failure decreases net promoter score by 3 pp." That said, don't overstate your results: if you're working with estimates, most likely you can round up your result to the nearest integer.

Key Takeaways

These are the key takeaways from this chapter:

Effective narratives in data science
Effective narratives are sequences of events connected by a story with the objective of driving action.

Properties of good narratives
To drive action, narratives must be clear and to the point, credible, memorable, and actionable.

Science as storytelling
I suggest an iterative approach to creating a narrative: start with the business problem and create stories or hypotheses that address the problem, test and refine them with the data iteratively, and finish with the delivery-stage narrative. This last narrative naturally evolves from the initial hypotheses.

Structure of a narrative
You may wish to follow the narrative arc: setup, struggle, and resolution. I've found it more effective to follow a simple and to-the-point storyline: what, so what, now what. These almost map one-to-one, but in data science I see little value in creating suspense or a sense of struggle.

TL;DRs and elevator pitches
These are great tools for achieving the right amount of simplification and to double-check that you indeed have a coherent narrative. TL;DRs may work as teasers for high-level executives who will only spend time going through the material if there's something memorable and actionable.

Practice makes perfect
Invest enough time practicing the delivery. If possible, record yourself.

Further Reading

There are many great references on narratives and storytelling with data. *Storytelling with Data* by Cole Nussbaumer Knaflic (Wiley) is great at improving your data visualization techniques, but also has a very good chapter on building narratives. I haven't covered datavis in this chapter, but this is a critical skill for data scientists who are creating a story. Chapter 8 goes into some of these skills. In a similar vein, *Effective Data Storytelling: How to Drive Change with Data, Narrative and Visuals* by Brent

Dykes (Wiley) is full of good insights. I found very useful his discussion on the interplay among data, visuals, and narrative.

Simply Said: Communicating Better at Work and Beyond by Jay Sullivan (Wiley) emphasizes the value of simplicity in general communication, written or not. His advice on writing short sentences (less than 10 words) is powerful.

It Was the Best of Sentences, It Was the Worst of Sentences: A Writer's Guide to Crafting Killer Sentences by June Casagrande (Ten Speed Press) is targeted at writers, but there are a ton of great suggestions to become a better communicators. Her emphasis on thinking about the audience ("the Reader is king") should be the North Star when building narratives.

Resonate: Present Visual Stories that Transform Audiences by Nancy Duarte (John Wiley and Sons) is great if you want to learn the art of storytelling from a designer's point of view. You will also find tons of details on many topics covered here in *Made to Stick: Why Some Ideas Survive and Others Die* by Chip Heath and Dan Heath (Random House). Their six principles will resonate quite a bit: simplicity, unexpectedness, concreteness, credibility, emotions, and stories.

The Hard Truth About Soft Skills: Workplace Lessons Smart People Wish They'd Learned Sooner by Peggy Klaus makes a strong case for focusing on your soft skills. Data scientists focus early on developing their technical expertise, thereby neglecting the so-called soft skills. The hard truth is that your career depends critically on the latter.

On narratives in the scientific method, the article by David Grandy and Barry Bickmore, "Science as Storytelling" (*https://oreil.ly/3rn9-*), provides more details on the analogy between the scientific method and storytelling.

## CHAPTER 8

Datavis: Choosing the Right Plot to Deliver a Message

Chapter 7 went through some good practices to build and deliver powerful narratives in data science. Data visualizations (datavis) are powerful tools to enrich your narratives and are a field of study on their own. As such, they need to be chosen as communication devices. The question you should always ask yourself is: *is this plot helping me convey the message I want to deliver?* If the answer is negative, you should go back to the drawing board and find the *right* plot for your message. This chapter goes through some recommendations that will help you improve your visualization skills.

Some Useful and Not-So-Used Data Visualizations

The field of datavis has evolved quite a bit in the last few decades. You can find online references, catalogues, and taxonomies that should help you find the right type of graph for your question. You can check the Data Visualisation Catalogue (*https://oreil.ly/BHQ1t*) or from Data to Viz (*https://oreil.ly/m75Ww*).

Unfortunately, many practitioners stick to default alternatives such as line and bar plots, often used interchangeably. In this chapter, I'll review some less well-known types of plots you can use, and discuss some pitfalls that are common among data practitioners. This is in no way exhaustive, so at the end of this chapter I'll point to some great resources that will provide a more complete picture of the field.

Bar Versus Line Plots

Let's start with the most basic question of all: when should you use bar and line plots? One common recommendation is to use bars for *categorical* data and lines for *continuous* data. The most common scenario for continuous data is when you have a time series, that is, a sequence of observations indexed by a time subscript (y_t). Let's check the validity of this recommendation.

Remember that a plot should help you deliver a message. With categorical data, such as average revenue per user across customer segments, most likely you want to highlight differences *across* segments. Moreover, there's no obvious ordering for the categories: you might want to sort them to help deliver a message or not, or you might just stick to alphabetical order. Bars are great communication devices since it's easy to look and compare the heights of the bars.

With time series, it's common to highlight several properties of the data:

- The sequential ordering that time provides
- The mean or average level
- The trend or growth rate
- Any curvature

Line plots are great if you care about any of these messages.

Figures 8-1, 8-2, and 8-3 display bars and lines for categorical data and two time series (short and long). Starting with categorical data, bars allow for easy comparison of the metric across segments. Line plots, on the other hand, are not great to visualize differences across segments. This is because the continuity of a line gives the incorrect perception that the segments are somehow connected. It requires some extra effort for the viewer to understand what you are plotting, putting at risk the message you want to deliver.

Looking at time series data, you might think that bars do an OK job, at least if the sample is short enough. Once you increase the sample size, unnecessary clutter arises, and then you must doubt your choice for the plot. Note that a line clearly and quickly tells you something about the trend and level, without the extra ink. More on this later when I discuss the data-ink ratio.

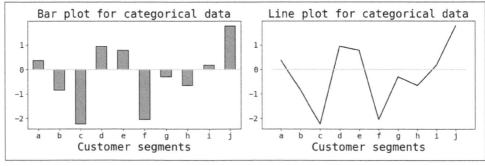

Figure 8-1. Bars and lines for customer segments

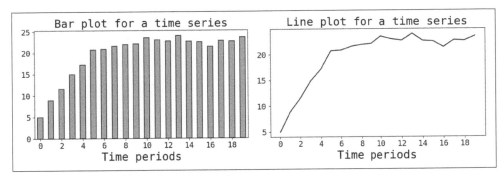

Figure 8-2. Bars and lines for a time series

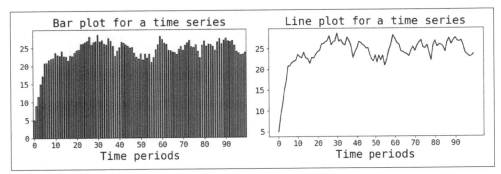

Figure 8-3. Bars and lines for a long time series

Slopegraphs

I learned about slopegraphs when I was going through Edward Tufte's *The Visual Display of Quantitative Information* (Graphics Press), and it took me some time to grasp their usefulness. One way to think about slopegraphs is that they are great when you need to convey a trend message for categorical data. In a sense, slopegraphs combine the best of bars and lines, since they allow you to compare trends across segments.

Figure 8-4 shows an example of a slopegraph. You can see that lines easily convey trends for each segment, and the visualization allows for easy comparisons across segments. In this example, I only have five segments, so getting the labels right is easy, but it can be challenging to get a readable plot with more segments. Nonetheless, there are other tools that can help you out with this, such as using a legend and different colors or line styles (such as dashes, dots, and the like).

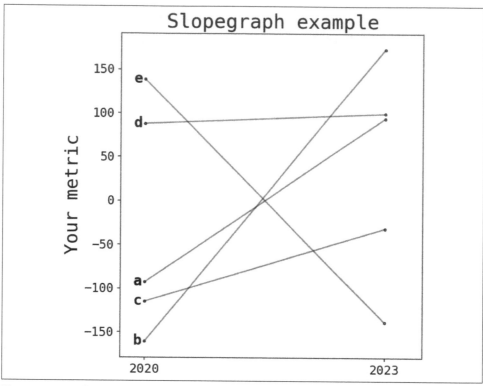

Figure 8-4. A slopegraph to highlight differences in trends

Waterfall Charts

Waterfall charts (Figure 8-5) are very often used by business stakeholders, and were famously popularized by McKinsey. The idea here is to *decompose* a change in a metric using segments or categories. I used waterfall charts in Chapter 3 since they're great at plotting the output of those decompositions.

Be careful when one of the segments has a substantially different scale, which often happens when you are using growth rates and some category has a very small starting value. Also, remember that this type of plot is useful when the message is about the decomposition.

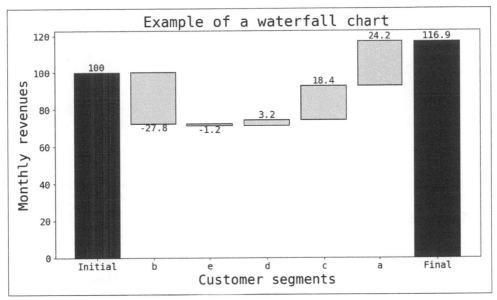

Figure 8-5. Revenue by customer segments

Scatterplot Smoothers

Scatterplots are great when you want to convey a message regarding the correlation between two variables X and Y. Unfortunately, with large datasets it's very hard to display that relationship, even if present.

There are several alternatives to handle this problem. The simplest solution is to create a plot with a random sample of your data. Generally, this is good enough since most of the time you don't need the complete dataset. Alternatively, you can use a hexagonal bin plot (*https://oreil.ly/sf_MH*) that in practice reduces the dimensionality by coloring the density of hexagonal areas. The same principle applies to contour plots (*https://oreil.ly/91Mtn*), but this requires a bit of preprocessing on your part.

An alternative solution is to use a *scatterplot smoother* that fits a nonlinear smoother on your data. This nonlinear function is general enough to help you find a relationship if there's one. You must be careful, however. One good principle in data visualization is to try not to alter the nature of the data (or *graphical integrity* as Tufte calls it), and smoothing techniques may alter the viewer's perception of the data.

Figure 8-6 shows three plots: the first scatterplot uses the entire dataset with 10 million observations. The second repeats the exercise with a small enough random sample of the original dataset. The third plot presents the original data and a cubic scatterplot smoother. Presenting the data is always a good practice: that way the viewers can decide for themselves if the smoother is a good representation of the relationship.

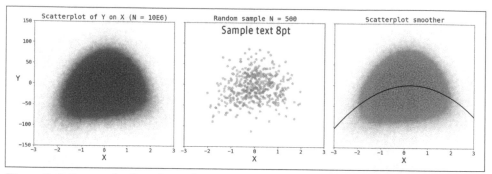

Figure 8-6. Scatterplot for a large dataset simulated with a quadratic generating process

Plotting Distributions

Distributions are critical for data scientists, and it's always a good practice to plot or print some quantiles of your metric *before* even starting to work on your project. It's less obvious that you should present distributions to your stakeholders, since they are hard to understand and this might create unnecessary confusion.

Histograms are the standard way to plot a distribution: these are just frequencies of occurrences in sorted, mutually exclusive subsets of the domain of your metric or bins. Kernel density estimates (*https://oreil.ly/29aJ3*) (KDE) plots give a smoothed estimate of the distribution and depend on two key parameters: a kernel or smoothing function and the bandwidth. Figure 8-7 shows a histogram and a Gaussian KDE for a simulated mixed-normal data.

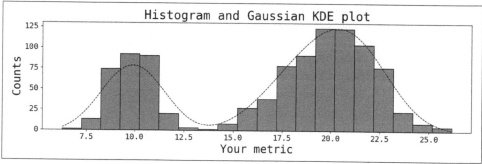

Figure 8-7. Histogram and KDE plot for simulated data

 When plotting KDEs, be careful with the scale. KDEs are smoothed estimates for the underlying distribution, ensuring that they integrate to one, rendering the scale meaningless. When I plot KDEs, I usually drop the vertical labels as they might create confusion. In Figure 8-7, I rescaled the axis to make it comparable to the one in the histogram.

With stakeholders, I rarely use histograms or KDEs as these usually have more information than needed to deliver the message. Most of the time you only need a few quantiles that can be presented with other visualizations, such as a standard *box plot* (*https://oreil.ly/mTEfe*). One exception is when I want to highlight something about the distribution that *matters to my message*; a typical use case is when I want to show that there's something in the domain of the metric that shows anomalous behavior, like in fraud prevention.

If you want to highlight *shifts* in the distribution, you can use box plots. A typical scenario is when you want to show that the *quality* of your sales or customers has changed, say because the average ticket has improved in time. Since the sample average is sensitive to outliers, you may want to show what has driven this change.

Figure 8-8 shows two alternative ways to plot these changes. The plot on the left shows the standard box plot, and on the right I just decided to plot the minimum and maximum, and the 25%, 50%, and 75% quantiles using a line plot. The box plot contains much more information than is needed to convey the message, so I decided to make two changes:

- Present only the data that is absolutely necessary (quantile).

- Use a line plot according to the recommendations at the beginning of the chapter.

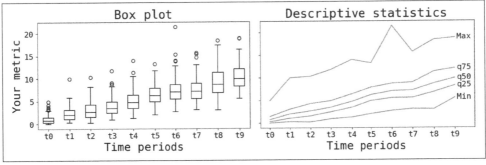

Figure 8-8. Two alternatives to plot changes in distribution

General Recommendations

After going through some common pitfalls in data visualization, let's jump right into some general recommendations for good design and execution.

Find the Right Datavis for Your Message

The type of plot you choose can alter the way your audience perceives the data, so you'd better find the one that really helps you convey your message. For instance, do you want to compare amounts between categories? A change in time? Proportions?

Uncertainty? You can find several resources online that will guide you, depending on what you want to communicate. For instance, the Data Visualisation Catalogue (*https://oreil.ly/_S7G-*) displays different types of plots depending on "what you want to show."

I can't reinforce enough that what matters is *the message*. As such, I always recommend trying several plots before deciding on the final output. It takes longer, but the last mile is critical. Figure 8-9 shows one plot that I discarded when preparing this chapter. It felt like a great idea to try a box plot and a line plot at the same time, but the delivery was not helping with my message (too much clutter).

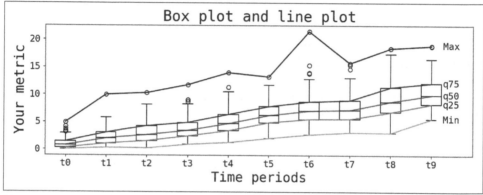

Figure 8-9. A plot that didn't help with the message

Choose Your Colors Wisely

One common mistake is thinking of color as a decorative feature for the plot. That may be true in a marketing setting, but in datavis, colors must be selected wisely to convey a message.

The common scenario is bar plots: you have one metric across categories and you want to show an interesting insight about one or several segments. A good recommendation is to choose *one and only one* color for all bars. I've seen many data science presentations where the speaker decides that the plot looks great if each bar has a different color. Take a step back and think about your viewers: will they think that you're superb at combining colors? That's possible, but many people actually think that the different colors represent a third variable that you want to highlight. In cases like this, where color conveys exactly the same information as the labels on the horizontal axis, it's better to choose only one color.

Figure 8-10 shows three examples: the first plot highlights what you want to *avoid*, since your segment labels and color represent the same dimension. The middle plot eliminates this redundancy. The third plot shows an example where coloring helps you deliver a message: you want your audience to focus on segment *b* that had terrible performance. If color is not enough, you can include other text annotations.

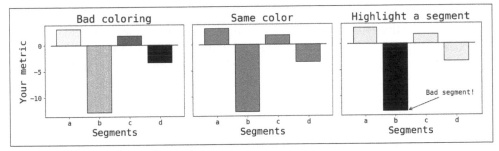

Figure 8-10. Examples with coloring

Different Dimensions in a Plot

The previous example can be generalized to other types of decorative features, such as different marker types or line styles. The same principle applies: use only one such feature if it conveys redundant information and may confuse the audience.

That said, you can use those extra features *if* you have additional information that is important for your message. The best example is a bubble plot: this is similar to a scatter plot where you want to say something about the relationship between two variables *X* and *Y*, and you also include a third variable *Z*, represented by the diameter of the circular marker, or bubble. An example is shown in Figure 8-11.

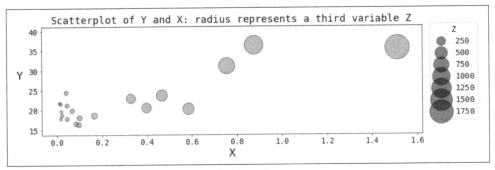

Figure 8-11. A third dimension with a bubble plot

Aim for a Large Enough Data-Ink Ratio

When discussing the use of bars in Figure 8-3, I mentioned that you should avoid clutter; the bars themselves are just providing redundant information. Edward Tufte formalized this intuition with the *data-ink ratio* concept. He defines *data-ink* as the "non-erasable core of a graphic." The *data-ink ratio* is the ratio of data-ink to the total ink in the graphic. You lower this ratio when you include noninformative features in the plot; alternatively, if you really just stick to representing the data and nothing else, you improve upon it.

While Tufte supports the idea of maximizing the ratio, I treat the data-ink ratio more as a North Star than a law set in stone, and there are indeed studies in visual perception that contradict Tufte's recommendation.[1] For instance, including extra information to highlight something for your audience, as in the rightmost plot in Figure 8-10, increases the ratio and is therefore a bad practice. You can judge for yourself, but I've found it useful to help the audience direct their attention to the details that help me get my point across.

Customization Versus Semiautomation

In a typical scenario, data scientists use visualization tools that help improve their productivity. These semiautomated tools reduce the time-to-delivery of the plot but usually provide little space for customization. Aim for flexible tools that allow you to easily customize your plots.

 I tend to support being on the customization side of this spectrum. Going *back to basics* with a general-enough tool like Python's Matplotlib that allows for a very large degree of customization can only improve your ability to create the right plot. The learning curve can be steep at the beginning, but after a while you'll be able to create almost any plot you imagine with no substantial effort.

Get the Font Size Right from the Beginning

This may sound like a no-brainer, but this is a mistake I see very frequently in data science presentations. Choose large enough font sizes for your plots and always check that every label is readable. And *always* include a title and labels for your vertical and horizontal axes. Aim at designing self-explanatory and readable plots.

A good practice when you use Python's Matplotlib is customizing the rcParams (*https://oreil.ly/m4rz3*). For instance, to ensure that I always have a large enough default font size, I always include something along the following lines at the top of my notebook or script, right after importing the necessary modules:

```
# set plotting parameters from the beginning
font = {'family' : 'monospace',
        'weight' : 'normal',
        'size'   : 14}
```

---

1 See, for example, McGurgan et al., "Graph Design: The Data-Ink Ratio and Expert Users," in *Proceedings of the 16th International Joint Conference on Computer Vision, Imaging and Computer Graphics Theory and Applications* (VISIGRAPP) 3 (2021): 188–194.

```
axes = { 'titlesize' : 22,
         'labelsize' : 20}
figure = {'figsize':(10,4),
          'autolayout':True}
matplotlib.rc('font', **font)
matplotlib.rc('axes', **axes)
matplotlib.rc('figure', **figure)
```

If you think that these new default parameters won't work for a specific graph, just overwrite them for that plot.

Interactive or Not

Interactive plots have gained quite a bit of popularity, first with the development of JavaScript libraries such as D3.js (*https://d3js.org*), and now with their availability in Python and R. In Python you can find several tools to make your plots interactive; among the most popular are Plotly (*https://plotly.com*), Seaborn (*https://oreil.ly/ CsVh7*), and Altair (*https://oreil.ly/zWKfz*), among others.

In static plots, such as the ones in this chapter, communication with the audience goes in one direction (from the creator to the audience). In many cases, this is not optimal since the audience can't *inspect the data* by themselves. Interactive plots help bridge this gap.

However, for most common use cases, they are just an overshoot. The recommendation here is to only use them whenever it's advisable that your audience inspect the data. Otherwise, stick to static plots with a clear message.

Stay Simple

In Chapter 7, I made a case for creating simple narratives, and this is especially true for data visualizations. Your objective is to deliver a message, and complex graphs make it unnecessarily difficult for your audience to understand it. Moreover, if you're giving a live presentation, it is very likely that you'll get questions that will distract you from your main message.

Start by Explaining the Plot

A common mistake is to assume that the audience understands the plot and jump right in to explain the main insights derived from it. That's why you should start by clarifying your plot: state clearly what's on the vertical and horizontal axis, and choose one part of the plot (such as a marker, line, or bar) and explain it. Once you ensure that the plot is clear, you can deliver your message.

Key Takeaways

These are the key takeaways from this chapter:

Purpose of data visualizations
Visualizations should aid you in delivering a message. Before presenting a plot, ensure that there is *a message* to deliver; otherwise, drop the plot.

Types of plots
Choose the type that best suits your delivery. Bars are great to compare a metric across categories. Lines are better if your metric is continuous or to display time series. Understand the difference and choose wisely.

General recommendations
Aim for simple visualizations and avoid clutter. Choose your colors wisely, and always ensure that the plot is readable by adjusting the font size. Ensure that your axes are labeled and that the labels make sense. Include the units when it's not self-explanatory. Avoid interactive plots unless strictly necessary.

Further Reading

One of the most cited references in datavis is *The Visual Display of Quantitative Information* by Edward Tufte (Graphics Press). Among many topics, he discusses in detail the *data-ink ratio*. Along with John Tukey and William Cleveland, Tufte is considered one of the foundational experts in the field.

Another mandatory reference for datavis enthusiasts is *The Grammar of Graphics* by Leland Wilkinson (Springer). R's popular *ggplot* library was inspired by Wilkinson's ideas, and it has had a profound effect on the profession and other widely used visualization libraries and tools.

A historical account of datavis can be found in "A Brief History of Data Visualization" (*https://oreil.ly/DcoGO*) by Michael Friendly, published in the *Handbook of Data Visualization* (Springer).

There are many great modern references on this topic. I highly recommend Claus Wilke's *Fundamentals of Data Visualization: A Primer on Making Informative and Compelling Figures* (O'Reilly).

Jake VanderPlas's *Python Data Science Handbook* (O'Reilly) has some great examples for topics discussed here and will help you understand some of the intricacies of Matplotlib. All of his code is on GitHub (*https://oreil.ly/Y698n*).

Kennedy Elliott's "39 Studies About Human Perception in 30 Minutes" (*https://oreil.ly/aneqb*) reviews some of the evidence regarding how different plots alter the perception of an object and their relative efficiency to convey distinct messages.

## PART II

Machine Learning

## CHAPTER 9

Simulation and Bootstrapping

The application of different techniques in the data scientist's toolkit depends critically on the nature of the data you're working with. *Observational* data arises in the normal, day-to-day, business-as-usual set of interactions at any company. In contrast, *experimental* data arises under well-designed experimental conditions, such as when you set up an A/B test. This type of data is most commonly used to infer causality or estimate the incrementality of a lever (Chapter 15).

A third type, *simulated* or *synthetic* data, is less well-known and occurs when a person re-creates the *data generating process* (DGP). This can be done either by making strong assumptions about it or by training a generative model on a dataset. In this chapter, I will only deal with the former type, but I'll recommend references at the end of this chapter if you're interested in the latter.

Simulation is a great tool for data scientists for different reasons:

Understanding an algorithm
> No algorithm works universally well across datasets. Simulation allows you to single out different aspects of a DGP and understand the sensitivity of the algorithm to changes. This is commonly done with Monte Carlo (MC) simulations.

Bootstrapping
> Many times you need to estimate the precision of an estimate without making distributional assumptions that simplify the calculations. Bootstrapping is a sort of simulation that can help you out in such cases.

Levers optimization
> There are some cases when you need to simulate a system to understand and optimize the impact of certain levers. This chapter doesn't discuss this topic, but does provide some references at the end.

Before delving into these topics, let's start with the basics of simulation.

Basics of Simulation

A *data generating process* (DGP) states clearly the relationships among inputs, noise, inputs, and outputs in a simulated dataset. Take this DGP as an example:

$$y = \alpha_0 + \alpha_1 x_1 + \alpha_2 x_2 + \epsilon$$
$$x_1, x_2 \sim N(\mathbf{0}, \Sigma)$$
$$\epsilon \sim N(0, \sigma^2)$$

This says that the dataset is comprised of one outcome (y) and two features (x_1, x_2). The outcome is a linear function of the features and noise. All of the information needed to simulate the dataset is included, so the DGP is fully specified.

To create the data, follow these steps:

1. *Set up some parameters.* Choose the values for $\alpha_0, \alpha_1, \alpha_2$, the 2×2 covariance matrix Σ, and the residual or pure noise variance σ^2.

2. *Draw from the distributions.* Here, I decided that the features follow a mean-zero multivariate normal distribution, and that the residuals are to be independently drawn from a mean-zero normal distribution.

3. *Compute the outcome.* Once all of the inputs are drawn, you can compute the outcome y.

The second step is at the heart of simulation, so let's discuss this first. Computers can't simulate truly random numbers. But there are ways to generate *pseudorandom* draws from distributions that have some desirable properties expected when there's pure uncertainty. In Python the random (*https://oreil.ly/rDWrf*) module includes several pseudorandom number generators that can be easily used.

Nonetheless, let's take a step back and try to understand how these pseudorandom number generators work. I will now describe the *inverse transform sampling* method.

Suppose you are able to draw a uniform random number $u \sim U(0, 1)$, and you want to draw from a distribution with a known cumulative distribution function (CDF), $F(x) = \text{Prob}(X \leq x)$. Importantly, you can also compute the inverse of the CDF. The steps are (Figure 9-1):

1. Generate K independent draws $u_k \sim U(0, 1)$.

2. For each u_k find $x_k = F^{-1}(u_k)$: the latter are independent draws from the desired distribution.

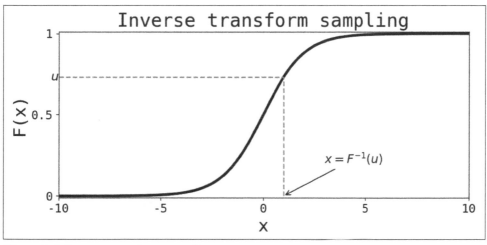

Figure 9-1. Inverse transform sampling

The following code snippet shows how to calculate the inverse CDF for a logistic random variable. Each uniform random number (*https://oreil.ly/yPHM6*) is passed as an argument, and then all you need to do is compute the inverse of the CDF, given some location and scale parameters:

```
def logistic_cdf_inverse(y, mu, sigma):
    """
    Return the inverse of the CDF of a logistic random variable
    Inputs:
        y: float: number between 0 and 1
        mu: location parameter
        sigma: scale parameter
    Returns:
        x: the inverse of F(y;mu,sigma)
    """
    inverse_cdf = mu + sigma*np.log(y/(1-y))
    return inverse_cdf
```

Figure 9-2 shows a Q-Q plot comparing Numpy's logistic random number generator (*https://oreil.ly/nnt5k*) and my own implementation using the inverse transform sampling just described, for three different sample sizes. Q-Q plots are great to visually inspect whether two distributions are similar. This is done by comparing corresponding quantiles for the distributions on the horizontal and vertical axes: equal distributions must have the same quantiles creating a plot that lies on the 45-degree diagonal (dashed), so you are looking for any departures from this ideal scenario. You can see that as the sample size increases, Numpy's logistic random number generator and my own implementation get closer.

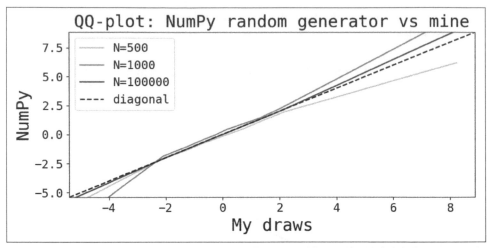

Figure 9-2. Numpy's and my own logistic random variable generator for different sample sizes

One last important piece of information has to do with the *seed* of random number generators. Pseudorandom numbers are generated through a dynamic process like $x_t = f(x_{t-1}, \cdots, x_{t-k}, x_0)$. The seed is the initial value of the sequence, so given the process (and its parameters), you can always replicate the complete sequence. In practice, seeds are used for the purpose of *replication*. In the code to this chapter, you'll see that I always set a seed to ensure that the results don't change when I run the code again.

Simulating a Linear Model and Linear Regression

The simplest simulation that is still useful in machine learning (ML) is that of a linear model. I will now simulate the following model:

$$y = 2 + 3.5x_1 - 5x_2 + \epsilon$$
$$x_1, x_2 \sim N(0, \mathbf{diag}(3, 10))$$
$$\epsilon \sim N(0, 1)$$

Note that the features are independent draws from a normal distribution (covariance matrix is diagonal, and bold denotes vectors or matrices), and that residuals follow a standard normal distribution.

You are now ready to run the MC simulation. A typical simulation is comprised of the following steps:

1. *Fix parameters, seeds, and sample size (N).* This ensures that one single MC experiment can be performed.

2. *Define what you wish to accomplish.* Typically, you want to test the performance of an ML algorithm against the true DGP, for instance, by computing the bias.

3. *Fix a number of simulations (M), estimate, and save the parameters.* For each experiment, simulate and train the model, and compute the metric defined in the previous step. For the case of bias, it would be something like:

$$\text{Bias}\left(\theta, \hat{\theta}\right) = E\left(\hat{\theta}\right) - \theta$$

Where θ is the *true* parameter of interest (set up in step 1), $\hat{\theta}$ is an estimate coming from an ML model, and the expectation is usually replaced with the sample mean across the M simulations.

Figure 9-3 shows the results from an MC simulation with three hundred experiments for the linear model defined and parameterized earlier. The estimated parameters for each experiment are saved, and the plot shows the sample mean and 95% confidence intervals, as well as the true parameters. The 95% confidence intervals are calculated directly from the results of the simulation by finding the 2.5% and 97.5% quantiles across the M experiments.

This is the *plain vanilla* simulation where all of the assumptions of ordinary least squares (OLS) are satisfied, so it's no surprise that linear regression does a great job at estimating the true parameters.

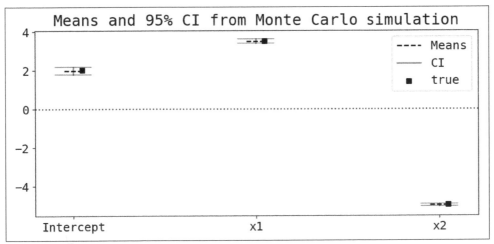

Figure 9-3. Results from an MC experiment with linear regression

Now that I've used an MC simulation to verify that OLS estimates are unbiased, I'll try something more interesting. What happens when the signal-to-noise ratio changes?

Intuitively, the *signal-to-noise ratio* (SNR) measures the amount of information provided by the model (signal) relative to that from the unexplained part of the model (noise). In general, the more informative features you include, the *higher* the SNR for your prediction model.

Using the first simulation as a baseline, it's straightforward to change the SNR by changing the residual variance σ^2 *and holding the variance of the features* fixed. Figure 9-4 plots the results from a new MC simulation with the same parameters as before, except for the residual variance, which is now one thousand times larger.

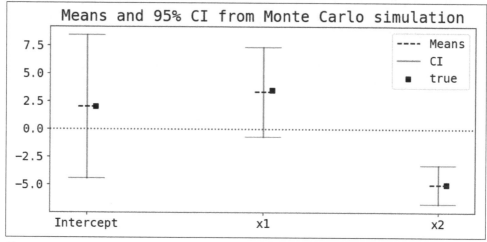

Figure 9-4. Linear regression and decreasing the SNR

You can visually validate that OLS remains unbiased in the sense that the average of the estimates is *very close* to the true parameters. But because of the lower SNR, estimates are now less precise (larger confidence intervals). This is a typical symptom when your SNR is not high enough.

Relevance of SNR

SNR is a very useful concept that all data scientists should feel comfortable with. Not so long ago, I was estimating the incrementality of a new product (B) on the company's revenue. This was particularly important because product B could be *cannibalizing* the older product (A), in the sense that the customers were not generating *more* revenue by using B, but rather they were substituting the use of A with B so that overall revenue remained the same.

This was a frustrating project because my team had already attempted incrementality estimation for the product, and had found mixed results (sometimes positive, sometimes negligible). I decided to use one of the techniques in Chapter 15 and again found that there was a positive but *statistically insignificant* effect. The reason was SNR: revenue from product B was still very low *relative* to the natural variance of the higher revenues from A. Learning this was humbling: even if there was an incremental effect, unless B scaled faster, you wouldn't be able to find it! Had I figured this out earlier, I could've saved a lot of time, effort, and organizational frustration.

What Are Partial Dependence Plots?

Notwithstanding its subpar predictive performance, linear regression is still great from an *interpretability* standpoint. To see this, take the simple linear model used before:

$$y = \alpha_0 + \alpha_1 x_1 + \alpha_2 x_2 + \epsilon$$

Since the residual is mean-zero by assumption, calculating the conditional expectation and partial derivatives, you get:

$$\frac{\partial E(y \mid \mathbf{X})}{\partial x_k} = \alpha_k$$

This shows that each parameter can be interpreted as the *marginal effect* of the corresponding feature on the expected outcome (conditioning on everything else). Put differently: in the linear world, a one-unit change in a feature is associated with a change in α_k units in the outcome. This makes OLS potentially great from a storytelling perspective.

Partial dependence plots (PDPs) are the counterpart for nonlinear models, such as random forest or gradient boosting regression:

$$y = f(x_1, x_2)$$

where f represents the possibly nonlinear function that you want to learn.

You can easily calculate PDPs for feature j by following these steps:[1]

1. *Train the model.* Train the model using the training sample, and save the model object.

2. *Calculate the means of the features.* Calculate the means of the K features $\bar{\mathbf{x}} = (\bar{x}_1, \cdots, \bar{x}_K)$. Because of random sampling, it shouldn't matter if you use the test or training sample.

3. *Create a linear grid for the j-th feature* x_j. Fix a grid size G and create the grid as $\text{grid}(x_j) = (x_{0j}, x_{1j}, \cdots, x_{Gj})$, where indices 0 and G are used to denote the minimum and maximum values of the feature in your sample.[2]

4. *Compute a means-grid matrix.* Matrix $\overline{\mathbf{X}}_j$ has the linear grid for x_j in the corresponding column, and means for all other features elsewhere:

$$\overline{\mathbf{X}}_j = \begin{pmatrix} \bar{x}_1 & \bar{x}_2 & \cdots & x_{0j} & \cdots & \bar{x}_K \\ \bar{x}_1 & \bar{x}_2 & \cdots & x_{1j} & \cdots & \bar{x}_K \\ \vdots & \vdots & \ddots & \vdots & \vdots \\ \bar{x}_1 & \bar{x}_2 & \cdots & x_{Gj} & \cdots & \bar{x}_K \end{pmatrix}_{G \times K}$$

5. *Make a prediction.* Using your trained model, make a prediction using the means-grid matrix. This gives you the PDP for feature j:

$$\text{PDP}(x_j) = \hat{f}(\overline{\mathbf{X}}_j)$$

Note that the *partial* derivative and the *partial* dependence plots answer a very similar question: what is the predicted change in the outcome when only one feature is allowed to vary? With nonlinear functions, you need to fix everything else at some value (standard practice is the sample mean, but you can choose otherwise). The partial derivative focuses on *changes*, while the PDP plots the entire predicted outcome given a feature that is allowed to vary.

---

1 While I find this method intuitive, it's not the standard way to compute PDPs. In Chapter 13, I'll discuss this in depth.

2 Alternatively, you can trim the outliers and set the extremes at some chosen quantiles. The code in the repo (*https://oreil.ly/dshp-repo*) allows for this setting, which is very helpful in applications.

The pseudocode I've shown you works well for continuous features. With categorical features, you need to be careful with the "grid": instead of creating a linear grid, you just create an array of possible values, such as {0,1} for a dummy variable. Everything else is the same, but a bar plot makes more sense here, as explained in Chapter 8.

I'll now use the first model I simulated to compare the results from a linear regression and scikit-learn's gradient boosting regression (*https://oreil.ly/UNDoi*) (GBR) and random forest regression (*https://oreil.ly/fFCoh*) (RFR). This is a useful benchmark to set: nonlinear algorithms are expected to be more powerful at identifying nonlinearities, but are they also good when the true underlying model is linear?

Figure 9-5 plots the true slopes along with the estimated PDPs for GBR and RFR, using the *maximum depth = 1* parameter that governs the maximum height of each tree in both algorithms. This is not an unreasonable choice here since the model is linear in the parameters and features; a single tree wouldn't be able to learn the DGP, but this restriction is less important for ensembles. All other metaparameters are fixed at scikit-learn's default values.

It's interesting to see that out-of-the-box GBR makes a great job at recovering the true parameter for both features. RFR does a decent job with x_2 but not with x_1.

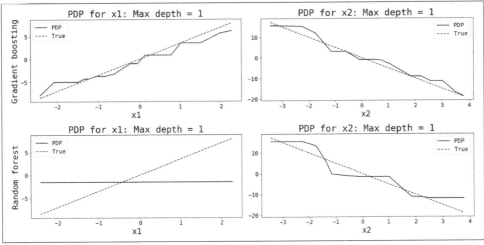

Figure 9-5. PDPs for GBR and RFR regression (max. depth = 1)

Figure 9-6 shows the results when *maximum depth* = 7 and everything else is set at the default values as before. GBR performs well again, and with the additional allowed nonlinearity, RFR is also able to estimate the true parameters. Interestingly, with maximum depth ≥ 3, the right shape of the PDP for x_1 starts to be recovered (results in the repo (*https://oreil.ly/dshp-repo*) for this chapter). What's going on here?

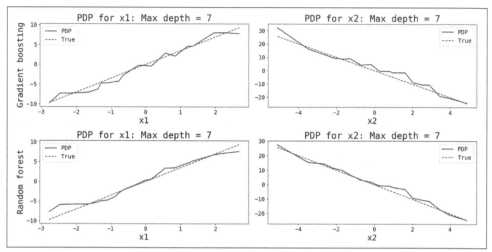

Figure 9-6. PDPs for GB and RF regression (max. depth = 7)

This simulation has two parameters that simultaneously give more weight to the second feature x_2: it was drawn from a normal distribution with higher variance ($\sigma_{22} = 10 > 3 = \sigma_{11}$), and the corresponding parameter is also larger in absolute value. This means that one standard deviation change in x_2 has a larger impact on y than a corresponding change in x_1. The result is that RFR tends to select the second feature more often in the first and only split of each tree.

Figure 9-7 shows the results when I switch the variances of the simulated features and everything else remains the same. You can see that RFR now does a better job at estimating the true effect of the first feature and a relatively bad one (but not terrible as before) for the second feature. Since the parameters were not changed—only the variance from the distributions that were drawn—x_2 still gets sufficient weight at the first splits of each tree in the ensemble, so the algorithm is able to pick up part of the true effect.

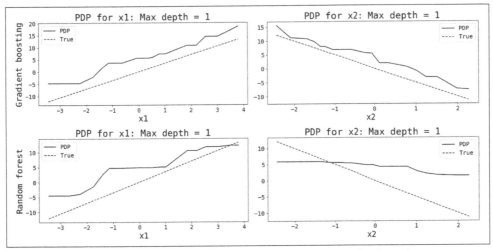

Figure 9-7. PDPs when variances of features are switched (max. depth = 1)

You may wonder if there's another metaparameter that can be optimized to reduce the bias in this RFR estimation. As mentioned, the problem appears to be that x_2 is given more weight, so it ends up being selected for the first split in the tree (and any further splits if the maximum depth is increased). One way to proceed is by changing the default parameter max_features that sets the number of *randomly* chosen features that are allowed to compete in each split. The default value is the total number of features (two in this example), so x_1 always loses. But if you change it to one feature, thanks to the randomness of the selection, you force the ensemble to give it a free pass sometimes. Figure 9-8 shows the results of making this change.

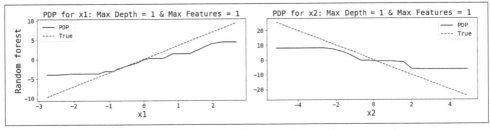

Figure 9-8. Random forest PDPs (max. depth = 1 and max. features = 1)

Omitted Variable Bias

In linear regression, *omitted variable bias* (*https://oreil.ly/IqzUA*) takes place when the data scientist fails to include one feature that *must have been included* and is correlated to any other included feature, creating biased parameter estimates and thus, suboptimal predictive performance.

To explain how the bias works, let's go back to the simple two-feature linear model presented at the beginning of the chapter, but assume now that the data scientist includes only the first feature and estimates:

$$y = \beta_0 + \beta_1 x_1 + \eta$$

The true unobserved coefficient for the included variable is α_1, so comparing it to the coefficient from the misspecified model (β_1), you can show that:

$$\beta_1 = \alpha_1 + \underbrace{\alpha_2 \frac{Cov(x_1, x_2)}{Var(x_1)}}_{Bias}$$

It follows that there's bias when the two features are uncorrelated. Moreover, the sign of the bias depends on the sign of $\alpha_2 \times Cov(x_1, x_2)$.

Let's start by simulating the same DGP as before, but excluding x_2. I'll do this for a grid of correlation coefficients, since these are bounded to the $[-1,1]$ interval and are thus easier to work with. Recall that the true parameter is $\alpha_2 = -5$, so the sign of the bias will be negative the sign of the correlation:

$$Sgn(Bias) = -Sgn(Cov(x_1, x_2))$$

To simulate $x_1, x_2 \sim N(0, \Sigma(\rho))$, you can simplify the parameterization by having unit variances so that:

$$\Sigma(\rho) = \begin{pmatrix} 1 & \rho \\ \rho & 1 \end{pmatrix}$$

The steps to run the simulation are as follows:

1. Fix a correlation parameter ρ from the grid.
2. Simulate the DGP given this correlation parameter.
3. For each MC experiment, estimate the parameters excluding the second feature.
4. Compute the bias across all MC experiments.
5. Repeat for all other elements of the grid.

Figure 9-9 shows the results from an MC simulation with different correlation parameters. Four results are noteworthy:

- Bias is null when features are uncorrelated.
- The sign of the bias is negative the sign of the correlation parameter.
- With unit correlation coefficient, bias equals the parameter of the excluded feature.
- There's no bias for the intercept (by definition, uncorrelated with the omitted variable).

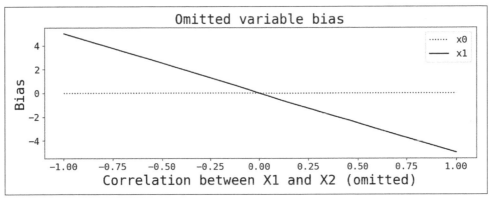

Figure 9-9. Bias as a function of the correlation parameter

Let's summarize this last finding: *if you're going to use linear regression, think really hard about the features you need to include!* That's why including some controls is always recommended even if you have only weak hypotheses about the underlying causal mechanism (for instance, including geographical dummies can help you mitigate the extent of omitted variable bias with features that vary at that level).

This said, these days almost no one uses OLS except in introductory courses or textbooks or when estimating causal effects (Chapter 15). A natural question is whether the more predictive algorithms also suffer from this problem.

To answer this, let's run an MC experiment and compute the bias from OLS and GBR. But I first need to find a way to estimate parameters with GBR that are comparable to those in the linear DGP. Inspecting a PDP (Figure 9-5) suggests a simple way to do it:

1. Construct the partial dependence plot for x_1.
2. Run a linear regression $pdp = \gamma_0 + \gamma_1 \text{Grid}(x_1) + \zeta$.
3. Use the estimated slope parameter γ_1 to compute the bias.

Figure 9-10 plots simulated bias for the case of independent (left plot) and correlated (right plot) features for OLS and GBR *without metaparameter optimization*. As expected, with independent features, bias is indistinguishable from zero (see the confidence intervals). With positively correlated features, bias is negative and statistically different from zero, and this is true for *both* OLS and GBR. The results are discouraging and humbling: you cannot fix a data problem with an algorithm.

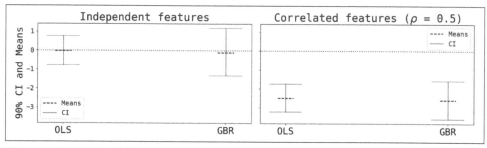

Figure 9-10. OLS and GBR bias for independent and correlated features

> As a general rule, don't expect your algorithm to fix a problem with your data. There are *robust* algorithms, but none is bulletproof.

Simulating Classification Problems

As you may recall, in classification problems the outcome variable is categorical rather than continuous. These problems arise frequently in data science (DS), with typical use cases that include predicting customer churn (two categories: a user churned or didn't churn), problems where a customer needs to accept or reject an offer, such as cross- and up-selling or any other marketing campaign, predicting fraud, and many others.

Latent Variable Models

One standard way to simulate binomial classification models is by using latent variables.[3] A variable is *latent* if it can't be directly observed but affects an observable outcome. This definition will become more clear after inspecting the following DGP:

---

3 To simulate a multinomial logistic model, you need to use a different technique (*https://oreil.ly/K5d8i*) that takes into account some properties of logistic multinomial models.

$$z = \alpha_0 + \alpha_1 x_1 + \alpha_2 x_2 + \epsilon$$

$$y = \begin{cases} 0 & \text{if } z < 0 \\ 1 & \text{if } z \geq 0 \end{cases}$$

$$\epsilon \sim \text{Logistic}(\mu, s)$$

$$x_1, x_2 \sim N(\mathbf{0}, \Sigma)$$

The latent variable is z, and it follows a simple linear model with logistic disturbances. You only observe the binomial variable y that depends on the sign of the latent variable.

The choice of the distribution for the disturbances can help you simulate models with more or less *balanced* outcomes. Symmetric distributions like the Gaussian or logistic generate balanced outcomes, but you can choose an asymmetric distribution if you want to focus the simulation on the "unbalancedness" of the data (you can also hand-pick different thresholds without changing the distribution and achieve the same result).

One important difference with linear regression models is that usually the parameters in the DGP for the latent variable are not *identifiable*, meaning that they can't be directly estimated; only *normalized* versions of the parameters can be estimated. To see this, notice that:

$$\begin{aligned} \text{Prob}(y = 1) &= \text{Prob}(\mathbf{x}'\alpha + \epsilon \geq 0) \\ &= \text{Prob}(-\epsilon \leq \mathbf{x}'\alpha) \\ &= \text{Prob}\left(-\frac{\epsilon}{\sigma_\epsilon} \leq \frac{\mathbf{x}'\alpha}{\sigma_\epsilon}\right) \\ &= F\left(\mathbf{x}'\alpha/\sigma_\epsilon\right) \end{aligned}$$

where F is the CDF for the logistic distribution, and I've used the fact that the logistic PDF is symmetric. The last equation shows that true parameters are indistinguishable from normalized parameters α/σ_ϵ. In the simulation, I will report both sets of parameters to highlight this fact.

Marginal effects in classification models measure the impact of an infinitesimal change in one feature on *the probability* of interest. In linear regression this was just the coefficient corresponding to each feature, but since CDFs are nonlinear in the parameters, the calculation is not as straightforward for classification models. Since the derivative of the CDF is the PDF, after applying the chain rule for differentiation, you get:

$$\frac{\partial \text{Prob}(y = 1)}{\partial x_k} = f(\mathbf{x}'\alpha)\alpha_k$$

$$= \frac{e^{\mathbf{x}'\alpha}}{\left(1 + e^{\mathbf{x}'\alpha}\right)^2}\ \alpha_k$$

Note how nonlinearity kicks in: in order to calculate the marginal effect of one feature, you need to evaluate $f(\mathbf{x}'\alpha)$. As with PDPs, the standard practice is to use the sample means of the features to compute the inner product with the estimated parameters. The sign of the marginal effect depends only on the sign of the true parameters, which is always a desirable property.

Comparing Different Algorithms

I will now run an MC simulation to compare the results from three different models:

Linear probability model
    Run OLS on the observed binary outcome and features. I do *not* correct for heteroskedasticity using weighted least squares, which is the standard practice when you want to report confidence intervals (but it doesn't affect the bias).[4]

Logistic model
    Standard logistic regression (*https://oreil.ly/rfsei*). I present both the estimated parameters and the marginal effects obtained from the last equation.

Gradient boosting classifier
    From the scikit-learn library (*https://oreil.ly/H3JkU*). To make it comparable, I compute the slope of the PDP.

The parameters for the simulation are as follows:

$$\left(\alpha_0, \alpha_1, \alpha_2\right) = (2, 3.5, -5)$$
$$\sigma_{11} = \sigma_{22} = s = 1$$
$$\sigma_{12} = \sigma_{21} = \mu = 0$$
$$\sigma_\epsilon^2 = \left(s^2 \pi^2\right)/3 \approx 3.28$$
$$\left(\alpha_0/\sigma_\epsilon, \alpha_1/\sigma_\epsilon, \alpha_2/\sigma_\epsilon\right) \approx (1.1, 1.9, -2.8)$$

---

[4] A critical assumption in OLS is that the disturbances have the same variance (*homoskedastic*). In contrast, *heteroskedastic* disturbances have different variance parameters, and OLS is no longer *optimal* in a very precise sense. Weighted least squares are an alternative to OLS when the form of the heteroskedasticity can be estimated.

The last line shows the true normalized parameters that will serve as a benchmark.

Results can be found in Figure 9-11. The two main lessons from this simulation are:

True parameters are not identified.
Compared to the true parameters in the DGP, estimated parameters from the logistic regression are off since they are not identifiable. Nonetheless, estimates are very close to *normalized parameters* as expected: compare the estimates (1.0, 1.8, −2.6) with the true normalized parameters earlier.

The three methods estimate the right marginal effects.
Theoretical marginal effects from the logistic regression (PDF times the coefficient), coefficients from a linear probability model, and PDP slopes from GBR are in agreement.

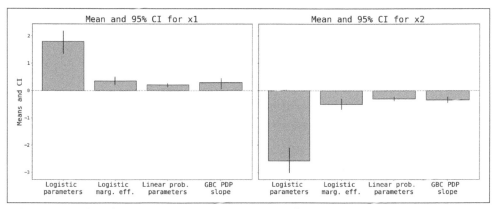

Figure 9-11. Classification simulation: comparison of estimates

---

Lessons from the Classification Simulation

When using a gradient boosting or random forest classifier, you can use the PDP to start opening the *black box*. In some cases, it's convenient to estimate linear regression directly on binary outcomes (linear probability model): estimated parameters can be interpreted as marginal effects on the *probability*. Be careful with confidence intervals and predicted values: to get *good* confidence intervals, you need to use weighted least squares or a *robust* estimator, and predicted values are not bounded to the unit interval (so you may end up with negative or larger-than-one probabilities).

Bootstrapping

Monte Carlo simulation is all about generating datasets by specifying the DGPs. In contrast, *bootstrapping* generates samples *from the current dataset* and is mostly used to quantify the variance of an estimate. Examples of estimates that are relevant in data science are PDPs (and marginal effects), precision and recall, and the like. Since these depend on the sample at your disposal, there will always be some sampling variation that you may want to quantify.

To describe how bootstrapping works, assume that the number of observations in your sample is N. Your estimate is a function of your sample data, so:

$$\hat{\theta} = \hat{\theta}\left(\{y_i, \mathbf{x_i}\}_{i=1}^{N}\right)$$

A pseudocode for bootstrapping is:

1. Set the number of bootstrap samples (B).
2. For each sample $b = 1, ..., B$:
   a. Randomly select, *with replacement*, N rows from your dataset.
   b. Calculate and save your estimate, given this bootstrap sample:

   $$\hat{\theta}^b = \hat{\theta}\left(\{y_i^b, \mathbf{x_i^b}\}_{i=1}^{N}\right)$$

3. Calculate the variance or confidence interval using the B estimates. For instance, you can calculate the standard deviation like this:

$$SD\left(\hat{\theta}\right) = \sqrt{\frac{\sum_{b=1}^{B}\left(\hat{\theta}^b - AVG\left(\hat{\theta}^b\right)\right)^2}{B-1}}$$

A typical use case is when you decide to plot the true positive rate (TPR) after dividing your sample in equally spaced buckets, such as deciles (see Chapter 6). In classification models it's natural to expect that the score is informative of the actual occurrence of the event, implying that the TPR should be a nondecreasing function of the score (higher scores, higher incidence).

To give a concrete example, suppose you trained a churn model predicting whether a customer will stop purchasing in the next month or not. You make a prediction for two customers and get the scores $\hat{s}_1 = 0.8$ and $\hat{s}_2 = 0.5$. Ideally, these would represent actual probabilities, but in most cases, scores and probabilities don't map one-to-one, so this requires some calibration. But even if the scores can't be interpreted as probabilities, it would be great if they're at least *directionally correct*, in the sense that the first customer is more likely to churn.

Plotting TPRs by buckets allows you to see if your model is informative in this sense. But there's a catch! Because of sampling variation, monotonicity really depends on the desired granularity. To see this principle in action, Figure 9-12 shows TPR for quintiles, deciles, and 20-tiles (ventiles), along with bootstrapped 95% confidence intervals. You can see that monotonicity holds when I use quintiles and deciles. What happens when you decide to increase the granularity to 20 equally spaced buckets? If you hadn't plotted the confidence intervals, you would've concluded that there's something off with your model (see buckets 11, 15, and 19). But it's all about sampling variation: once you take this into account, you can safely conclude that these buckets are not statistically different from their neighbors.

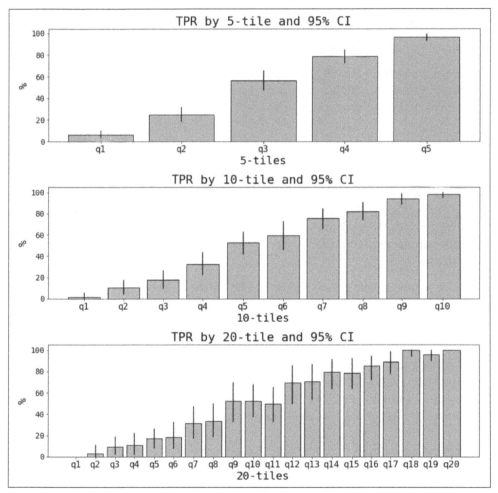

Figure 9-12. Bootstrapping TPRs from a classification model

If you have a statistics background, you may think that bootstrapping is unnecessarily complicated in this example, since you just need to calculate the parametric variance of the TPR per bucket, which follows a binomial distribution (so for deciles, the variance can be calculated as $N/10 \times TPR_d \times (1 - TPR_d)$). With this you can calculate parametric confidence intervals. You're right; bootstrapping is most useful when:

- You want to calculate the variance without making distributional assumptions (that is, nonparametric estimation).

- Computing the variance analytically is hard or computationally expensive.

Key Takeaways

These are the key takeaways from this chapter:

No algorithm works universally well across datasets.
Since real-world data is not perfect, you may want to check if the algorithm performs correctly in a simulated example.

Algorithms won't fix data issues.
Knowing the limitations of each training algorithm is critical. Moreover, if you have issues with your data, don't expect the algorithm to fix them.

Simulation as a tool for understanding algorithm limitations.
In this chapter, I've presented several examples where simulation provides insights into the pros and cons of different algorithms. Other examples can be found in the repo (*https://oreil.ly/dshp-repo*) for this chapter (outliers and missing values).

Partial dependence plots are great tools for opening the black box of many ML algorithms.
To showcase the power of simulation, I computed PDPs and compared them to the parameters of linear regression and classification.

Bootstrapping can help you quantify the precision of your estimates.
Bootstrapping is similar to Monte Carlo simulation in the sense that you draw repeated samples—not from simulated DGPs but from your dataset—and infer some statistical properties with this information.

Further Reading

The field of simulation is vast, and this chapter barely scratched the most basic principles. Simulation is an essential tool in Bayesian statistics and generative ML models. For the former, you can check Andrew Gelman et al., *Bayesian Data Analysis*, 3rd ed. (Chapman and Hall/CRC Press). A great reference for the latter is Kevin Murphy's *Machine Learning: A Probabilistic Perspective* (MIT Press). He also has two updated versions that I haven't reviewed but should be great.

Monte Carlo Statistical Methods by Christian Robert and George Casella (Springer) is a now classic reference on the vast and complex field of Monte Carlo simulations and how to draw from distributions. Note that this book is for the technically inclined.

You can find more information on bootstrapping in *The Elements of Statistical Learning: Data Mining, Inference, and Prediction*, 2nd ed., by Trevor Hastie et al. (Springer, and available online (*https://oreil.ly/QvSUb*) on the author's web page). You can also find information on some of the methods used for linear and logistic regression.

Practical Synthetic Data Generation by Khaled El Emam et al. (O'Reilly) provides some useful information on simulating synthetic data. As I mentioned at the beginning of the chapter, you can simulate data by making assumptions about the data generating processes behind a dataset, or you can train a model on real-world data that can be used to generate synthetic datasets. This book provides some practical guidance on how to do it.

The omitted variable bias and the lack of identification in logistic regression are pretty standard results that can be found in any econometrics textbook. See for instance William Greene's *Econometric Analysis*, 8th ed. (Pearson).

In *Analytical Skills for AI and Data Science*, I discuss the use of simulation for levers optimization. Scott Page's *The Model Thinker: What You Need to Know to Make Data Work for You* (Basic Books) is a good reference if you want to explore this subject. See also *Stochastic Simulation* by Brian Ripley (Wiley).

## CHAPTER 10

Linear Regression: Going Back to Basics

Linear regression (OLS[1]) is the first machine learning algorithm most data scientists learn, but it has become more of an intellectual curiosity with the advent of more powerful nonlinear alternatives, like gradient boosting regression. Because of this, many practitioners don't know many properties of OLS that are very helpful to gain some intuition about learning algorithms. This chapter goes through some of these important properties and highlights their significance.

What's in a Coefficient?

Let's start with the simplest setting with only one feature:

$$y = \alpha_0 + \alpha_1 x_1 + \epsilon$$

The first parameter is the *constant* or *intercept*, and the second parameter is the *slope*, as you may recall from the typical functional form for a line.

Since the residuals are mean zero, by taking partial derivatives you can see that:

$$\alpha_1 = \frac{\partial E(y)}{\partial x_1}$$

$$\alpha_0 = E(y) - \alpha_1 E(x_1)$$

---

1 OLS stands for ordinary least squares, which is the standard method used to train linear regression. For convenience I treat them as equivalent, but bear in mind that there are other loss functions that can be used.

As discussed in Chapter 9, the first equation is quite useful for interpretability reasons, since it says that a one-unit change in the feature is associated with a change in α_1 units of the outcome, on average. However, as I will now show, you must be careful not to give it a *causal* interpretation.

By substituting the definition of the outcome inside the covariance, you can also show that:

$$\alpha_1 = \frac{Cov(y, x_1)}{Var(x_1)}$$

In a bivariate setting, the slope depends on the covariance between the outcome and the feature, and the variance of the feature. Since correlation is not causation, you must be cautious not to interpret these *causally*. A non-null covariance can arise from different factors:

Direct causation
> As you would like to interpret it $(x_1 \to y)$.

Reverse causation
> $x_1 \leftarrow y$, since the covariance is symmetric on the arguments.

Confounders
> A confounder is any third variable that affects both x and y, but these are otherwise unrelated (Figure 10-1).

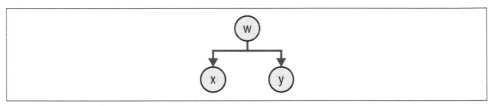

Figure 10-1. Confounders

 Estimates from linear regression provide information about the degree of correlation between a feature and an outcome, and it can only be interpreted *causally* in very specific situations (see Chapter 15). This warning also applies to other ML algorithms such as gradient boosting or random forests.

A more general result applies for multiple regression (that is, a regression with multiple covariates):

$$\alpha_k = \frac{Cov(y, \tilde{x}_k)}{Var(\tilde{x}_k)}$$

where $\tilde{x}_k$ is the residual from running a regression of the *k-th* feature on all *other* features (−k):

$$\tilde{x}_k = x_k - \mathbf{X}_{-k}\theta_{-k}$$

For the bivariate linear model, the snippet in Example 10-1 shows that linear regression and the simpler covariance formula agree numerically.

Example 10-1. Verifying that OLS and the bivariate covariance formula agree

```
def compute_alpha_1feat(ydf, xdf):
    """Compute alpha using var-cov formula and linear regression
        for the simple case of y = a + b x
    Inputs:
        ydf, xdf: dataframes with outcome and feature
    Outputs:
        Estimated coefficients from two methods: Cov() formula and
        linear regression
    """
    # Using covariance formula
    cat_mat = ydf.copy()
    cat_mat['x'] = xdf['x1']   # concatenate [y|x] so I can use the .cov() method
    vcv = cat_mat.cov()
    cov_xy = vcv['y'].loc['x']
    var_x  = vcv['x'].loc['x']
    beta_vcv = cov_xy/var_x
    # using linear regression
    reg = LinearRegression(fit_intercept=True).fit(xdf, ydf.values.flatten())
    beta_reg = reg.coef_[0]

    return beta_vcv, beta_reg

# compute and print
b_vcv, b_reg = compute_alpha_1feat(ydf=ydf, xdf=Xdf[['x1']])
decimals = 10
print(f'Alpha vcv formula = {b_vcv.round(decimals=decimals)}')
print(f'Alpha OLS = {b_reg.round(decimals=decimals)}')

Alpha vcv formula = 3.531180168,
Alpha OLS = 3.531180168
```

For the case of more than one feature, you can use the following function to verify that the more general covariance formula agrees with OLS. Note that I first compute the residuals of a regression of feature k on all other features:

```
def compute_alpha_n_feats(ydf, xdf, name_var):
    """

    Compute linear regression coefficients by:
        1. Orthogonalization (Cov formula)
        2. OLS
    Inputs:
        ydf, xdf: dataframes with outcome and features
        name_var: string: name of feature you want to compute
    Outputs:
        Coefficient for name_var using both methods

    """
    # Run regression of name_var on all other features and save residuals
    cols_exc_x = np.array(list(set(xdf.columns) - set([name_var])))
    new_x = xdf[cols_exc_x]
    new_y = xdf[name_var]
    reg_x = LinearRegression().fit(new_x, new_y.values.flatten())
    resids_x = new_y - reg_x.predict(new_x)
    # Pass residuals to Cov formula
    cat_mat = ydf.copy()
    cat_mat['x'] = resids_x
    vcv = cat_mat.cov()
    cov_xy = vcv['y'].loc['x']
    var_x  = vcv['x'].loc['x']
    beta_vcv = cov_xy/var_x
    # using linear regression
    reg = LinearRegression().fit(xdf, ydf.values.flatten())
    all_betas = reg.coef_
    ix_var = np.where(xdf.columns == name_var)
    beta_reg = all_betas[ix_var][0]

    return beta_vcv, beta_reg
```

The more general covariance formula leads to an important result called the *Frisch-Waugh-Lovell theorem*.

The Frisch-Waugh-Lovell Theorem

The Frisch-Waugh-Lovell theorem (FWL) is a powerful result that helps build a lot of intuition about the inner workings of linear regression. It essentially says that you can interpret the OLS estimates as *partialled-out* effects, that is, effects net of any other dependencies between features.

Say that you're running a regression of sales per customer on the price they paid and state dummy variables. If a stakeholder asks you if the price coefficient can be explained by state-wise variation in pricing, you can use the FWL theorem to convincingly say that these are *net effects*. The price effect has already been cleaned out of any differences in pricing across states (you have already *controlled* for state differences).

To present the theorem I'll use the simpler two-feature linear model again, but the theorem applies to the more general case of any number of regressors:

$$y = \alpha_0 + \alpha_1 x_1 + \alpha_2 x_2 + \epsilon$$

FWL states that you can estimate a specific coefficient, say α_1, by using a two-step process:

1. *Partialling out x_2*:

    a. Run a regression of y on x_2 and save the residuals: $\tilde{y}_1$.

    b. Run a regression of x_1 on x_2 and save the residuals: $\tilde{x}_1$.

2. *Regression on residuals*:

    a. Run a regression of $\tilde{y}_1$ on $\tilde{x}_1$. The slope is an estimate of α_1.

The partialling-out step removes the effect of any other regressor on the outcome and the feature of interest. The second step runs a bivariate regression on these residuals. Since we have already partialled out the effect of x_2, only the effect of interest remains.

Example 10-2 shows the results when I simulate a linear model with three features, and estimate each coefficient using the FWL *partialling-out* method and plain linear regression. I use the code snippet in Example 10-2 to make the comparison.

Example 10-2. Checking the validity of FWL

```
def check_fw(ydf, xdf, var_name, version = 'residuals'):
    """
    Check the Frisch-Waugh theorem:
        Method 1: two-step regressions on partialled-out regressions
        Method 2: one-step regression
    Inputs:
        ydf, xdf: dataframes with Y and X respectively
        var_name: string: name of feature we want to apply the FW for
        version: string: ['residuals','direct'] can be used to test
            both covariance formulas presented in the chapter
            'residuals': Cov(tilde{y}, tilde{x})
            'direct': Cov(y, tilde{x})
    """
    # METHOD 1: two-step regressions
    nobs = ydf.shape[0]
    cols_exc_k = np.array(list(set(xdf.columns) - set([var_name])))
    x_k = xdf[cols_exc_k]
    # reg 1:
    reg_y = LinearRegression().fit(x_k, ydf.values.flatten())
    res_yk = ydf.values.flatten() - reg_y.predict(x_k)
    # reg 2:
```

```
new_y = xdf[var_name]
reg_x = LinearRegression().fit(x_k, new_y.values.flatten())
res_xk = new_y.values.flatten() - reg_x.predict(x_k)
res_xk = res_xk.reshape((nobs,1))
# reg 3:
if version=='residuals':
    reg_res = LinearRegression().fit(res_xk, res_yk)
else:
    reg_res = LinearRegression().fit(res_xk, ydf.values.flatten())
coef_fw = reg_res.coef_[0]
# METHOD 2: OLS directly
reg = LinearRegression().fit(xdf, ydf.values.flatten())
coef_all = reg.coef_
ix_var = np.where(xdf.columns == var_name)[0][0]
coef_ols = coef_all[ix_var]

return coef_fw, coef_ols


cols_to_include = set(Xdf.columns)-set(['x0'])
decimals= 5
print('Printing the results from OLS and FW two-step methods \nVersion = residuals')
for col in ['x1', 'x2', 'x3']:
    a, b = check_fw(ydf, xdf=Xdf[cols_to_include], var_name=col, version='residuals')
    print(f'{col}: FW two-steps = {a.round(decimals=decimals)},
        OLS = {b.round(decimals=decimals)}')

Printing the results from OLS and FW two-step methods
Version = residuals
x1: FW two-steps = 3.66436, OLS = 3.66436
x2: FW two-steps = -1.8564, OLS = -1.8564
x3: FW two-steps = 2.95345, OLS = 2.95345
```

Going back to the covariance formula presented earlier, FWL implies that:

$$\alpha_k = \frac{Cov(\tilde{y}_k, \tilde{x}_k)}{Var(\tilde{x}_k)}$$

where as before, $\tilde{x}_k$ denotes the residuals from a regression of feature k on all other features, and $\tilde{y}_k$ denotes the residuals from a regression of the outcome on the same set of features. The Python script allows you to test that both versions of the general covariance formula give the same results (using the `version` argument).

An important property of OLS is that the estimated residuals are orthogonal to the regressors (or any function of the regressors), a process also known as *orthogonalization*. You can use this fact to show that the two covariance formulas are equivalent.

Importantly, orthogonalization *always* has to be performed on the feature of interest. If you only orthogonalize the outcome y, the covariance formula is no longer valid, *unless* the features are already orthogonal with each other, so in general:

$$\alpha_k \neq \frac{Cov(\tilde{y}_k, x_k)}{Var(x_k)}$$

Why Should You Care About FWL?

I've presented several versions of the orthogonalization result, so you should expect it to be relevant. The main takeaway is this:

> You can interpret each coefficient from linear regression as the net effect of each feature *after* cleaning it from the effects from any other feature.

Here's one typical scenario where this interpretation matters a lot:

$$x_1 \sim N\left(0, \sigma_1^2\right)$$
$$x_2 = \beta_0 + \beta_1 x_1 + \epsilon$$
$$y = \alpha_0 + \alpha_1 x_1 + \alpha_2 x_2 + \eta$$

In this case, x_1 has a direct and an indirect effect on the outcome y. An example could be your state or geographic dummy variables. These tend to have direct and indirect effects. When you interpret the coefficient of x_2, it would be great if you can say that this is *net* of any state differences, since you're already *controlling* for that variable.

Figure 10-2 shows the true parameter, OLS estimate, and gradient boosting regression (GBR) partial dependence plot (PDP) for a simulation of the previous data generating process. Thanks to FWL, you know that OLS will capture net effects correctly. GBR does well for x_2, but not so well for x_1.

To understand what's going on, recall how PDPs are calculated: fix one feature at the sample mean, create a grid for the one you care about, and make a prediction. When you fix x_2, x_1 displays a combination of direct and indirect effects, and the algorithm doesn't know how to separate them. This just reinforces the message that OLS is great for interpretability purposes, but requires quite a bit of effort to get the performance that even a relatively out-of-the-box GBR has with nonlinear models.

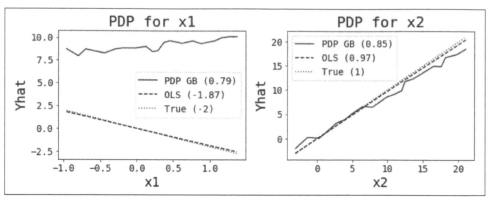

Figure 10-2. OLS and gradient boosting with direct and indirect effects

Confounders

Now that I've described the FWL theorem, I want to go back to the problem of confounders (Figure 10-1). Suppose that a confounder (w) affects two otherwise unrelated variables:

$$x = \alpha_x + \beta_x w + \epsilon_x$$
$$y = \alpha_y + \beta_y w + \epsilon_y$$
$$\epsilon_x \perp\!\!\!\perp \epsilon_y$$
$$\epsilon_x, \epsilon_y \perp\!\!\!\perp w$$

where the symbol $\perp\!\!\!\perp$ denotes statistical independence. Using the covariance formula for the slope coefficient in a regression of y on x, it becomes apparent why OLS shows spurious results:

$$\frac{Cov(y, x)}{Var(x)} = \frac{\beta_x \beta_y Var(w)}{\beta_x^2 Var(w) + Var(\epsilon_x)}$$

What if you first *cleaned* that common factor out? That's exactly what FWL tells you that linear regression does, so you can safely run a regression of the form:

$$y = \alpha_0 + \alpha_1 x_1 + \alpha_2 w + \epsilon$$

By also including the common factor w, OLS will effectively partial out its effect. Figure 10-3 shows the results of estimating the bivariate and spurious regression (left plot) and the partialled-out version when you also include the third factor as in the previous equation (right plot). I also include 95% confidence intervals.

Without controlling for the confounder, you would conclude that x and y are indeed correlated (confidence interval away from zero), but once you control for w, this becomes the only relevant (statistically significant) factor.

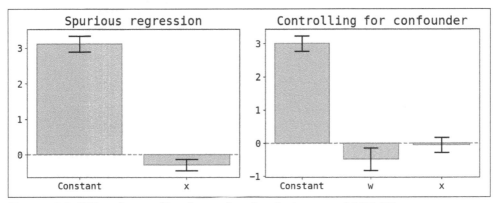

Figure 10-3. FW and controlling for confounders (estimate and 95% CI)

This result is very useful in many applications. In time series analysis, for example, it's quite common to have *trend-stationary* (*https://oreil.ly/ewcVV*) variables that can be modelled like this:

$$y_{1t} = \alpha_1 + \beta_1 t + \epsilon_{1t}$$
$$y_{2t} = \alpha_2 + \beta_2 t + \epsilon_{2t}$$

Thanks to FWL, you already know why these are called *trend-stationary*: once you control for a time trend (t above), thereby cleaning them from this effect, you end up with a stationary time series.[2]

Suppose you run a regression of one on the other:

$$y_{2t} = \theta_0 + \theta_1 y_{1t} + \zeta_t$$

Since you're not controlling for the common trend, you will end up incorrectly concluding that they are correlated. Figure 10-4 shows regression results from a

---

2 At a high level, a time series is *stationary* when its probability distribution doesn't change in time. Weak stationarity refers only to the first two moments, and strong stationarity requires that the joint distribution is constant. The mean for a trending variable changes, so it can't be stationary (unless it's *trend-stationary*).

simulation of two trend-stationary AR(1) processes that are unrelated by design.[3] The plot shows the estimated intercept (*constant*) and slope for the second variable (*y2*), as well as 95% confidence intervals.

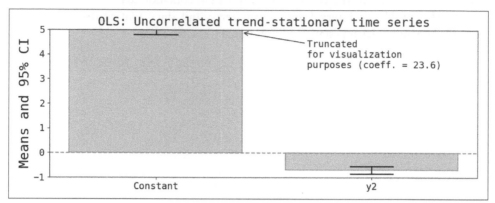

Figure 10-4. *OLS on spurious time series regression*

It's quite common to have spurious correlation with time series, as they most often display a time trend. Since it can act as a confounder, it's always recommended to include a linear time trend as a control. This way you *clean up* any noise that may arise from this potential confounder.

Additional Variables

Chapter 9 described the *omitted variable bias* that showed that *excluding* a variable that should have been included results in biased OLS estimates and thus, reduced predictive performance; importantly, this is also true for other machine learning (ML) algorithms.

What happens if instead of omitting important variables, you include additional irrelevant features? One of the nice properties of OLS is that including uninformative features creates no bias, and only affects the variance of the estimates. Figure 10-5 reports the mean and 90% confidence intervals for each estimated parameter from a Monte Carlo simulation, where:

- Only one feature is informative (x_1, with true coefficient $\alpha_1 = 3$).

- Four more uninformative controls are included when the model is trained.

---

3 *AR(1)* denotes an autoregressive process of order 1.

- Two models are trained: OLS and an out-of-the-box gradient boosting regression.

Both algorithms perform correctly on two fronts: they are able to correctly estimate the true parameter, and dismiss the uninformative variables.

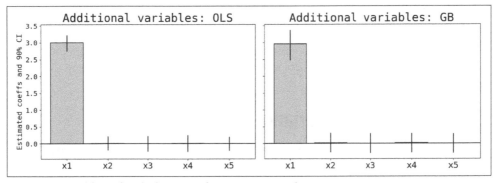

Figure 10-5. Effect of including uninformative controls

However, you must be cautious with ensemble learning algorithms since these tend to be quite sensitive when uninformative features are included, if these are highly correlated with the real underlying variables. You can typically see this with the *dummy variable trap*. The typical scenario arises with models with a dummy variable, like the following:

$$y = \alpha_0 + \alpha_1 x + \alpha_2 D_l + \epsilon$$

$$D_l = \begin{cases} 1 & \text{if} \quad \text{customer is left-handed} \\ 0 & \text{if} \quad \text{customer is right-handed} \end{cases}$$

In OLS, the dummy variable trap arises when you include an intercept *and* dummies for *all* available categories. In this example, you can only include *one* dummy variable for left- *or* right-handedness, but not both, because the cross-product matrix $X'X$ is not invertible (and thus the OLS estimates don't exist).[4] The solution is to always leave out the dummy variable for a *reference category*, in the example, the right-handed category.

This computational restriction doesn't exist with ensemble algorithms like random forests or gradient boosting regression, but since dummy variables like D_l and $D_r = 1 - D_l$ are perfectly correlated, it's normal to find both ranking very high in terms of feature importance. Since they provide the *exact same information*, the performance

---

4 Recall that the OLS estimator is $(X'X)^{-1}X'Y$.

of the algorithm doesn't improve by including both. This is one useful intuitive fact that arises naturally by understanding OLS.

Avoid the Dummy Variable Trap

In OLS, there is a dummy variable trap when you include an intercept and dummies for all categories in a categorical variable; in this case, the estimator doesn't exist.

If you use ensemble learning, you don't have this computational restriction, but these redundant features provide no extra information or predictive performance.

The Central Role of Variance in ML

One central tenet in ML is that you need variation in the features *and* the outcome for your algorithm to *identify* the parameters, or put differently, to learn the correlation. You can see this directly in the covariance formulation presented at the beginning: if x or y are constant, the covariance is zero, and hence OLS can't learn the parameter. Moreover, if x is constant, the denominator is zero, and thus the parameter doesn't exist, a result strongly connected to the dummy variable trap.

You need variation in the inputs if you want to explain variation in the output. This is true for any ML algorithm.

You may recall that in OLS, the estimates for the coefficients and the covariance matrix are:

$$\hat{\beta} = (X'X)^{-1}X'Y$$
$$\mathrm{Var}\left(\hat{\beta}\right) = s^2(X'X)^{-1}$$

where s^2 is the sample estimate of the residual variance, and $X_{N \times P}$ is the feature matrix, including the vector of ones that correspond to the intercept.

From these equations, two results follow:

Conditions for identification
> There can't be perfect correlation between features (perfect multicollinearity) for the cross-product matrix $(X'X)$ to be positive definite (full rank or invertible).

Variance of the estimates
> The more correlated the features, the higher the variance of the estimates.

While the first part should be straightforward, the second requires a bit of mathematical manipulation to show for the general case of multiple regression. In the repo (*https://oreil.ly/dshp-repo*) for this chapter, I include a simulation that verifies this condition in the case of multiple regression. For a simple bivariate regression, it's easy to show that the variance of the estimate is *negatively related* to the sample variance of the feature, so having covariates that exhibit more variation provides more information, thereby improving the precision of the estimates.[5]

Figure 10-6 plots the average and 95% confidence intervals for the estimates from OLS and gradient boosting regression after simulating a bivariate linear DGP where $Var(x_1)$ is increased over a grid. As discussed, for OLS the variance of the estimate *decreases* as the covariate displays *more* variation. Notably, the same is true for GBR.

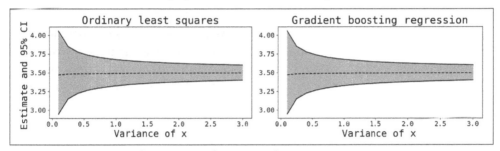

Figure 10-6. Variance of an estimate for OLS and GBR

This principle is at play in a practice that is not uncommon among data scientists. Imagine that you're running a regression like the following:

$$y_i = \alpha + \sum_s \theta_s D_{is} + \gamma \bar{z}_{s(i)} + \epsilon_i$$

$$D_{is} = \begin{cases} 1 & \text{if customer } i \text{ lives in state } s \\ 0 & \text{otherwise} \end{cases}$$

$\bar{z}_{s(i)}$ = state sample average of z given the state where i lives

If y denotes sales per customer and z household income, this model says that sales vary across states (dummy variables) and that there's an independent effect whereby richer states also purchase more (proxied with the average household income for each state).

While your intuition might be right, you won't be able to train this model with OLS since there's perfect multicollinearity. In other words, the state dummy variables and

---

5 In a bivariate setting, $Var(\beta_1) = Var(residual)/Var(x)$.

state averages *of any metric you can think of* provide the *exact same information*. And this is true for any ML algorithm!

Dummy Variables and Group-Level Aggregates

If you include dummy variables to control for group-level variation, there's no need to include aggregates for any other feature at that same level: the two provide exactly the same variation.

For instance, if you include state dummy variables, and you're tempted to also include *average household spend per state* and *median prices per state*, no matter how different they sound, they provide exactly the same amount of information.

To check this, I simulate a simple model using the data generating process I just covered, where I include three states (and thus, two dummy variables to avoid the dummy variable trap) drawn from a multinomial distribution (code can be found in the repo (*https://oreil.ly/dshp-repo*)). Example 10-3 shows that the features matrix is indeed low rank, implying that there's perfect multicollinearity.

Example 10-3. State dummies: effect of dropping the state average

```
# Show that X is not full column rank (and thus, won't be invertible)
print(f'Columns of X = {Xdf.columns.values}')
rank_x = LA.matrix_rank(Xdf)
nvars = Xdf.shape[1]
print(f'X: Rank = {rank_x}, total columns = {nvars}')
# what happens if we drop the means?
X_nm = Xdf[[col for col in Xdf.columns if col != 'mean_z']]
rank_xnm = LA.matrix_rank(X_nm)
nvars_nm = X_nm.shape[1]
print(f'X_[-meanz]: Rank = {rank_xnm}, total columns = {nvars_nm}')

Columns of X = ['x0' 'x1' 'D1' 'D2' 'mean_z']
X: Rank = 4, total columns = 5
X_[-meanz]: Rank = 4, total columns = 4
```

To check that the same point is valid for more general nonlinear algorithms, I ran a Monte Carlo (MC) simulation of this same model, training it with gradient boosting regression (no metaparameter optimization) and calculated the mean squared error (MSE) for the test sample using the complete set of features and after dropping the redundant mean feature. Figure 10-7 shows the average MSE along with 90% confidence intervals for MSE. You can verify that the predictive performance is virtually the same, as you would expect if the extra variable provides no additional information.

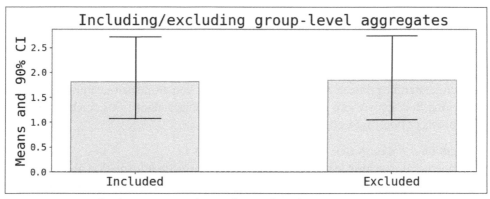

Figure 10-7. Results from MC simulation for gradient boosting

Use Case: Detecting Fraud

To see the power of this intuition, let's go into a high-stakes use case. If the goal is to build a fraud-detection ML model, which features should you include? Here's one possible story that applies very generally.

Since fraudsters don't want to get caught, they will try to appear as *normal* consumers. However, they don't usually know the distribution of the metrics, whereas you *do* know it. Suppose you have some metric x, like the amount or ticket of the transaction. One way you can transform it to help the algorithm detect an anomaly is by creating a ratio of the feature relative to some benchmark, like the 95% quantile:

$$x_{norm} = \frac{x}{x_{q95}}$$

With this transformation, whenever x is higher than the selected quantile, it will be larger than one, and the hope is that the algorithm might detect some pattern indicative of fraud.

By now you'd be right to be suspicious about this logic, since the normalized feature has *exactly the same information* as the original one. In the code repo (*https://oreil.ly/dshp-repo*) you can find an MC simulation for this use case, and you can verify that this is indeed correct.

Key Takeaways

These are the key takeaways from this chapter:

Why learn linear regression?
Understanding linear regression should help you build some important intuitions that apply more generally to other nonlinear algorithms, such as random forests or boosting techniques.

Correlation is not causation.
In general, machine learning algorithms only provide information about the correlation of features and outcome. The result is clear cut in linear regression, so this should serve as your benchmark when thinking about other learning algorithms.

Frisch-Waugh-Lovell theorem.
This is an important result in linear regression that states that the estimates can be interpreted as the net effect after controlling for the remaining covariates.

FWL and confounders.
Thanks to FWL, you can control for confounders just by including them in your set of features. One common example is in time series analysis, where it's always a good practice to control for a deterministic trend. This acts as a safeguard against getting spurious results when the outcome and features display some trend.

Irrelevant variables.
In linear regression, it is safe to include uninformative controls. Ensemble learning algorithms might be sensitive to irrelevant variables if these are sufficiently correlated to informative features. You won't bias your estimates, but this may lead you to conclude that some variable has predictive power when it doesn't.

The dummy variable trap.
In linear regression, it's always a good practice to include an intercept or constant term. If you include dummy variables, you must always exclude one category that will serve as a reference or base. For instance, if you include a female dummy variable, the male category serves as a reference for interpretation purposes.

The dummy variable trap in ensemble learning.
Nothing forbids you from including dummy variables for all categories with random forest or gradient boosting machines. But you also gain *nothing* from it: these variables provide no extra information that can improve the predictive performance of your model.

Variance is critical for machine learning.

Without sufficient variance in your features, your algorithm won't be able to learn the underlying data generating process. This is true for linear regression and general machine learning algorithms.

Further Reading

Linear regression is covered in most statistics, machine learning, and econometrics textbooks. The treatment in Trevor Hastie et al., *The Elements of Statistical Learning: Data Mining, Inference, and Prediction*, 2nd ed. (Springer), is superb. It discusses *regression by successive orthogonalization*, a result that is closely related to the FWL theorem.

Chapter 3 of *Mostly Harmless Econometrics: An Empiricist's Companion* by Joshua Angrist and Jörn-Steffen Pischke (Princeton University Press) provides a very deep discussion on the fundamentals of linear regression, as well as the derivations for the covariance formulas presented in the chapter. This book is great if you want to strengthen your intuitions on regression.

The FWL theorem is covered in most econometrics textbooks. You can check out William Greene's *Econometric Analysis*, 8th ed. (Pearson).

## CHAPTER 11

Data Leakage

In "Leakage in Data Mining: Formulation, Detection, and Avoidance," Shachar Kaufman et al. (2012) identify data leakage as one of the top 10 most common problems in data science. In my experience, it should rank even higher: if you have trained enough real-life models, it's unlikely you haven't encountered it.

This chapter is devoted to discussing data leakage, some symptoms, and what can be done about it.

What Is Data Leakage?

As the name suggests, *data leakage* occurs when some of the data used for training a model isn't available when you deploy your model into production, creating subpar predictive performance in the latter stage. This usually happens when you train a model:

- Using data or metadata that won't be available at the prediction stage
- That is correlated with the outcome you want to predict
- That creates *unrealistically high* test-sample predictive performance

The last item explains why leakage is a source of concern and frustration for data scientists: when you train a model, absent any data and model drift, you expect that the predictive performance on the test sample will extrapolate to the real world once you deploy the model in production. This won't be the case if you have data leakage, and you (your stakeholders and the company) will suffer a big disappointment.

Let's go through several examples to clarify this definition.

Outcome Is Also a Feature

This is a trivial example, but helps as a benchmark for more realistic examples. If you train a model like this:

$$y = f(y)$$

you'll get perfect performance at the training stage, but needless to say, you won't be able to make a prediction when your model is deployed in production (since the outcome is, by definition, not available at the time of prediction).

A Function of the Outcome Is Itself a Feature

A more realistic example is when one of the features is a function of the outcome. Suppose you want to make a prediction of next month's revenue and, using the $P \times Q$ decomposition described in Chapter 2, you include the unit price (Revenue/Sales) as a feature. Many times, the unit price calculation is done upstream, so you just end up using a table that contains prices without really knowing how they are calculated.

The Importance of Data Governance

The case of a feature that is itself a function of the outcome highlights the importance of data governance for data science, and machine learning more specifically. Having well-documented pipelines, thorough data lineage tracking, and variable definitions is a critical asset in any data-driven company.

Data governance may be costly for the organization, but the returns from embarking on it early on are well worth it.

Bad Controls

As described in Chapter 10, it's good to include features that you may think help control for sources of variation, even if you don't have a strong hypothesis for the underlying causal mechanism. This is generally true, unless you include *bad controls*, which are themselves outcomes affected by the features.

Take these data generating processes (DGPs) as an example:

$$y_t = f(\mathbf{x}_{t-1}) + \epsilon_t$$
$$z_t = g(y_t) + \zeta_t$$

You may think that controlling for z when training a model to predict y can help you clean out some of the effects. Unfortunately, since z won't be available at the time of

---

prediction, and is correlated with y, you end up with a nontrivial example of data leakage.

Note that leakage here arises both from using information that's not present at the time of prediction *and* from including the bad control. If z displays enough autocorrelation in time, even if you control for its lagged value (z_{t-1}), you will still have unreasonably high predictive performance.

Mislabeling of a Timestamp

Suppose you want to measure the number of monthly active users in a given month. A typical query that would produce the desired metric looks like this:

```
SELECT DATE_TRUNC('month', purchase_ts) AS month_p,
    COUNT(DISTINCT customer_id) AS mau
FROM my_fact_table
GROUP BY 1
ORDER BY 1;
```

Here you have effectively labeled these customers using the beginning-of-month timestamp, which for many purposes might make sense. Alternatively, you could've labeled them using the end-of-period timestamp, which could also be appropriate for different use cases.

The point is that the labeling choice may create data leakage if you incorrectly think that the metric was measured *before* the time suggested by your timestamp (so you would, in practice, be using information from the *future* to predict the *past*). This is a common problem encountered in practice.

Multiple Datasets with Sloppy Time Aggregations

Suppose you want to predict customer churn using a model like this:

$$\text{Prob}(churn_t) = f\left(\Delta\text{sales}_{t-1}^t, \text{num. products}_t\right)$$

There are two hypotheses at work here:

- Customers who have decreased their sales in the previous period are more likely to churn (they are effectively signaling their decreased engagement).
- Customers with a deeper relationship with the company, as measured by the number of other products they are currently using, are less likely to churn.

One possible cause for leakage occurs when the second feature includes information from the future, so that trivially, a customer who is active with one product next month *cannot* have churned. This might occur because you end up querying your data with something like the following code:

```
WITH sales AS (
-- subquery with info for each customer, sales and delta sales,
-- using time window 1
  ),
prods AS (
 -- subquery with number of products per customer using time window 2
 )
SELECT sales.*, prods.*
FROM sales
LEFT JOIN prods ON sales.customer_id = prods.customer_id
AND sales.month = prods.month
```

The problem arises because the data scientist was sloppy when filtering the dates in each subquery.

Leakage of Other Information

The previous examples dealt with leakage of data, either from the features or the outcome itself. In the definition, I also allowed for *metadata* leakage. This next example will help clarify what this means. In many ML applications, it's normal to transform your data by standardizing it like this:

$$y_{std} = \frac{y - \text{mean}(y)}{\text{std}(y)}$$

Suppose you standardize your *training* sample using the moments from the *complete* dataset, which of course includes the *test* sample. There are cases where these leaked moments provide extra information that won't be available in production. I'll provide an example later in this chapter that showcases this type of leakage.

Detecting Data Leakage

If your model has *unreasonably superior* predictive performance, you should suspect that there's data leakage. Not so long ago, a data scientist from my team was presenting the results from a classification model that had an area under the curve (AUC) of 1! You may recall that the AUC is bounded between 0 and 1, where an AUC = 1 means that you have a perfect prediction. This was clearly suspicious, to say the least.

These extreme cases of having a perfect prediction are quite rare. In classification settings, I get suspicious whenever I get an AUC > 0.8, but you shouldn't take this as a law written in stone. It's more of a personal heuristic that I've found useful and informative with the class of problems I have encountered in my career.[1] In regression settings it's harder to come up with similar heuristics, since the most common performance metric, the mean square error, is bounded from below by zero, but it really depends on the scale of your outcome.[2]

Ultimately, the best way to detect leakage is by comparing the real-life performance of a productive model with the test sample performance. If the latter is considerably larger, and you can rule out model or data drift, then you should look for sources of data leakage.

 Use your and your organization's knowledge of the modeling problem at hand to decide what is a suspicious level of superior predictive performance. Many times, detecting data leakage only happens when you deploy a model in production and get subpar performance relative to that of your test sample.

To show the improved performance from data leakage, I ran Monte Carlo (MC) simulations for two of the examples described earlier. Figure 11-1 shows the impact of including a bad control: I train models with and without data leakage, and the plot shows the mean and 90% confidence intervals across MC simulations. The mean squared error (MSE) with leakage is around a quarter of when the bad control is not included. With the code in the repo (*https://oreil.ly/hi693*), you can check that when the *bad control* is independent from the outcome, there's no data leakage and the models have the same performance. You can also tweak the degree of autocorrelation to check that even a lagged bad control can create leakage.

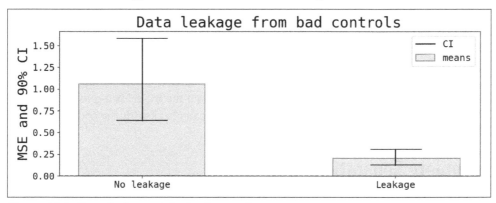

Figure 11-1. Data leakage with a bad control

---

1 Also, remember that the AUC is sensitive to having imbalanced outcomes, so my heuristic really must be taken with a grain of salt, to say the least.

2 An alternative is to use the coefficient of determination or *R2* that is also bounded to the unit interval.

In the second example, I'll show how bad standardization and leaking moments can affect the performance. Figure 11-2 presents mean MSE as well as 90% confidence intervals from an MC simulation using the following DGP:[3]

$$x_t \sim AR(1) \text{ with a trend}$$

$$y_t = f(x_t) + \epsilon_t$$

I use the first half of the sample for training purposes and the second half to test the model. For the *leakage* condition, I standardize the features and outcome using the complete dataset mean and standard deviation; for the *no leakage* condition, I use the moments for each corresponding sample (train and test). As before, it's quite manifest how the data leakage artificially improves the performance.

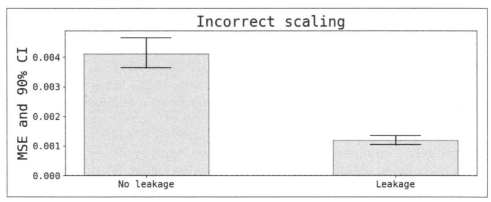

Figure 11-2. Data leakage from incorrect scaling (MSE)

What's the logic behind this type of leakage problem? I decided to include a time trend so that the mean and standard deviation from the complete dataset informs the algorithm *at training time* that the outcome and feature are increasing, thereby providing extra information that won't be available when the model is deployed. Without a trend component the leakage disappears, as you can check with the code in the repo (*https://oreil.ly/hi693*).

Complete Separation

Before moving on, I want to discuss the topic of *complete or quasi-complete separation*. In classification models, you may have an unusually high AUC because of this phenomenon, and this may or may not signal data leakage.

---

3 An *AR(1)* process is a time series with an autoregressive component of order 1. You can check Chapter 10 for more information.

Complete separation arises in the context of linear classification (think logistic regression) when a linear combination of the features perfectly predicts the outcome *y*. In cases like these, the minimum loss function (many times, the negative of the log likelihood function) doesn't exist. This typically happens when the dataset is small, when you work with imbalanced data, or when you used a continuous variable and a threshold to create a categorical outcome *and* include the variable as a feature. In the latter case, there is data leakage.

Quasi-complete separation arises when a linear combination of the features perfectly predicts a *subset* of your observations. This is much more common, and may happen when you include one or several dummy variables that, when combined, create a subset of observations for which there's a perfect prediction. In this case, you may need to check if there's data leakage or not. For instance, it may be that there's a business rule that says that cross-selling can only be offered to customers who live in a given state and if they have a minimum tenure. If you include tenure and state dummy variables, you will have quasi-complete separation and data leakage.

Real-Life Battle Scars from Quasi-Complete Separation

Several years ago a data scientist from my team was presenting the results from a classification model designed to improve the company's cross-selling campaign efficiency. The predictive performance was not incredibly high, but given his choice of features, I did find it unreasonably high.

When I asked him to open the black box, we found that the top variable in terms of predictive performance was a state dummy variable, which made no sense at all for this use case (there was nothing about the product that made it a better fit for customers in that state). After discussing the results with the sales team, we quickly realized that in the past quarter, the cross-selling campaigns had only targeted customers from that state. As a matter of fact, the sales team geographically rotated the campaigns to avoid being identified by the competitors. Since the data scientist had included data from the past two quarters, the state dummy created quasi-complete separation.

Many people are reluctant to label this as data leakage since the state dummies *will be* available at prediction time. They argue that it's more likely a case of *model drift* since part of the DGP appears to change in time. I prefer to label it as metadata leakage that can be easily avoided by excluding the state dummies, since the DGP doesn't really change. (The underlying factors for a customer to accept or reject an offer are the same *given that they get an offer*. But they have to get an offer!)

Let's simulate a case where this happens by using a latent variable approach, as described in Chapter 9. The data generating process is as follows:

$$x_1, x_2 \sim N(0, 1)$$

$$z = \alpha_0 + \alpha_1 x_1 + \alpha_2 x_2 + \epsilon$$

$$y = \mathbf{1}_{z \geq 0}$$

$$x_{3i} = \begin{cases} 1 & \text{for } i \text{ rand. selected from } \{j : y_j = 1\} \text{ with probability } p \\ 0 & \text{otherwise} \end{cases}$$

where $\mathbf{1}_{z \geq 0}$ is an indicator variable that takes the value 1 when the condition on the subscript applies and 0 otherwise.

The idea is simple: the true DGP is a binomial latent variable model with two covariates, but I create a third feature, used at training time, by randomly selecting without replacement from the observations with $y_i = 1$. This way I can simulate different degrees of separation, including the case of complete and no separation ($p = 1$ and $p = 0$, respectively). As usual, I train a logistic regression and a gradient boosting classifier (GBC) without metaparameter optimization.

I ran an MC simulation, and Figure 11-3 plots the lift in median AUC on the test sample across all experiments, where I benchmark everything with respect to the case of no separation. You can see that separation creates an increased AUC of up to 10% to 15% relative to the baseline, depending on whether I use a logistic regression or GBC.

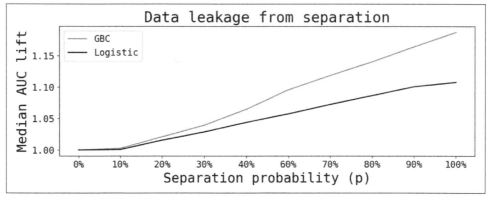

Figure 11-3. Lift in AUC for quasi-complete separation

The lesson here is that separation increases the AUC in a classification setting, and this may indicate data leakage that needs to be inspected further.

Windowing Methodology

I will now describe a windowing methodology that should help reduce the likelihood of data leakage in your model. As described earlier, data leakage can occur for many different reasons, so this is in no way a bulletproof technique. Nonetheless, I've found that it helps you discipline the process of training a model, and reduces some of the most obvious risks of leakage.

As a starting point, I separate the learning process into two stages:

Training stage
> This is the stage where you train a model by dividing your sample into training and testing, etc.

Scoring stage
> Once you've trained your model and you've deployed it in production, you use it to score a sample. It can be a one-at-a-time prediction, as in online, real-time scoring, or scoring of a larger sample.

It's easy to forget, but in machine learning (ML), *the scoring stage reigns*. It's so important that I devote Chapter 12 to discuss some necessary properties and processes that need to be set up to ensure that this stage is at its best. For now, just remember that this stage is where most value is created, and since this should be your North Star, it should be granted nobility status.

In ML, the scoring stage takes the leading role, and everything else should be set up to maximize the quality and timeliness of the predictions of this stage.

Figure 11-4 shows how the window methodology works. The starting point is the scoring stage (downmost timeline). Suppose you want to make a prediction at time t_p. This time period serves to divide the world into two windows:

Prediction window
> You're usually interested in predicting an event or a random variable associated to an event. For this you need to set up a prediction window for that event to occur $(t_p, t_p + P]$. For example, you want to predict if a customer will churn *in the next 30 days*. Or you want to predict your company's revenue *during the first quarter of the year*. Or you may want to predict if a customer will rate a book or a movie in *the next two weeks after* finishing reading or watching it.

Observation window
> Once you define a time horizon for your prediction to be evaluated, you need to define how much history you want to include to inform your prediction $[t_p - O, t_p]$. The name is derived from the fact that this, and only this, is the information we *observe* at scoring time.

Note that the prediction window is *open* on the left by design: this helps prevent data leakage as it explicitly separates what you observe at the time of prediction.

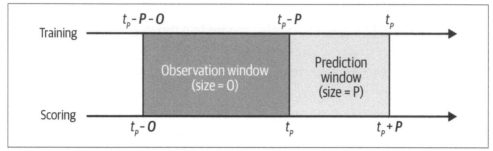

Figure 11-4. Windowing methodology

Let's go through an example to ensure that these concepts are clear. I want to train a churn model that predicts for each customer the likelihood they will churn during the next month. Since the scoring stage reigns, suppose I want to score all of the active users *today* (t_p). By definition, the prediction window starts from tomorrow and ends one month from tomorrow. At that point I have to be able to assess whether any of the customers churned or not. To make this prediction, I will use the last three months of information, so this is my observation window. Any transformations of the features are restricted to this timeframe. For instance, I may think that the most recent past matters, so I can compute the ratio of weekly interactions four weeks ago to last week's (if the ratio is larger than one, engagement increased this last month).

Choosing the Length of the Windows

You may wonder who chooses the lengths of the observation and prediction windows, and what considerations are taken into account. Table 11-1 summarizes some of the main considerations when deciding the lengths of both windows.

Table 11-1. Considerations when choosing window lengths

	Prediction (P)	Observation (O)
Owner	Business - data scientist	Data scientist
Predictive performance	Feasibility of short- vs. long-term prediction	Relative weight of distant past
Data	Historical data at your disposal	Historical data at your disposal

The length of the observation window is chosen by the data scientist, primarily based on the predictive performance of the model. For instance, is the recent past more predictive? The prediction window is primarily chosen by taking into account business considerations regarding the timeliness of a decision, and as such, it ought to be mostly owned by the business stakeholder.

It's important to acknowledge that longer prediction windows are usually less risky, in the sense that it's harder to be wrong (for example, predicting the existence of artificial general intelligence in the next one thousand years versus the next two years). But really short time horizons may be infeasible given the current granularity of your data (for example, predicting if a customer will churn in the next 10 minutes when you only have daily data).

Finally, the length of the prediction window affects how long the observation window must be. If the CFO asks me to predict revenue over the next five years, I can either have short observation windows and dynamic forecasts (where forecasts are used successively as features), or I can have a long enough observation window to make such a heroic forecast.

The Training Stage Mirrors the Scoring Stage

Once these windows are defined at the scoring stage, you're now ready to set up and define the training stage. As you might suspect from Figure 11-4, the training stage should always mirror what happens in the later scoring stage: the observation and prediction windows for the training stage map one-to-one to those at the scoring stage, and are thus constrained by them.

For instance, it's quite customary that you want to train the model with the most recent data at your disposal. Since you'll need P time periods to evaluate your prediction and O periods to create your features, this means that you need to set $[t_p - P - O, t_p - P]$ as your training observation window, and $(t_p - P, t_p]$ as your training prediction window.

Formally defining these windows helps you discipline and constrain what you expect to accomplish with the model. On the one hand, it ensures that only historical data is used for future predictions, preventing common leakage issues. You can see this more directly in the following equations:

$$\text{Scoring} \quad : \quad y_{\left(t_p, t_p + P\right]} = f\left(X_{\left[t_p - O, t_p\right]}\right)$$

$$\text{Training} \quad : \quad y_{\left(t_p - P, t_p\right]} = f\left(X_{\left[t_p - P - O, t_p - P\right]}\right)$$

Implementing the Windowing Methodology

Once you have defined them, you can enforce these on your code with something like the following snippet:

```
import datetime
from dateutil.relativedelta import relativedelta
```

```python
def query_data(len_obs: int, len_pre: int):
    """
    Function to query the data enforcing the chosen time windows.
    Requires a connection to the company's database

    Args:
        len_obs (int): Length in months for observation window (O).
        len_pre (int): Length in months for prediction window (P).

    Returns:
        df: Pandas DataFrame with data for training the model.
    """
    # set the time variables
    today = datetime.datetime.today()
    base_time = today - relativedelta(months = len_pre)  # t_p - P
    init_time = base_time - relativedelta(months = len_obs)
    end_time = base_time + relativedelta(months = len_pre)

    init_str = init_time.strftime('%Y-%m-%d')
    base_str = base_time.strftime('%Y-%m-%d')
    end_str = end_time.strftime('%Y-%m-%d')

    # print to check that things make sense
    print(f'Observation window (O={len_obs}): [{init_str}, {base_str})')
    print(f'Prediction window (P={len_pre}): [{base_str}, {end_str}]')
    # create query
    my_query = f"""
      SELECT
          SUM(CASE WHEN date >= '{init_str}' AND date < '{base_str}'
          THEN x_metric ELSE 0 END) AS my_feature,
          SUM(CASE WHEN date >= '{base_str}' AND date <= '{end_str}'
          THEN y_metric ELSE 0 END) AS my_outcome
      FROM my_table
    """
    print(my_query)
    # connect to database and bring in the data
    # will throw an error since the method doesn't exist
    df = connect_to_database(my_query, conn_parameters)
    return df
```

Summing up, the window methodology helps you enforce a minimal requirement that you can only use the past to predict the future. Other causes of data leakage may still be present.

I Have Leakage: Now What?

Once you have detected the source of leakage, the solution is to remove it and retrain your model. In some cases this is quite obvious, but in others it can take substantial time and effort. Here's a list of things you can attempt to identify the source of leakage:

Check time windows.

Ensure that you're always using past information to predict the future. This can be done by enforcing a strict time windowing process such as the one just described.

Check any data transformations done and enforce best practices.

A good practice is to use scikit-learn pipelines (*https://oreil.ly/iOEs1*) or similar to ensure that the transformations are done with the correct datasets and that there are no leaking moments or metadata.

Ensure that you have thorough knowledge of the business processes behind the creation of the data.

The more you know about the processes behind the creation of your data, the easier it is to identify potential sources of leakage or quasi-complete separation in the case of classification models.

Iteratively remove features.

On a regular basis you should run a diagnostic check to identify the most predictive features (in some algorithms you can do this with feature importance (*https://oreil.ly/uW6PY*)). Coupled with your knowledge of the business, this should help you identify if something looks *off*. You can also try iteratively removing the most important features to see if predictive performance changes dramatically in any iteration.

Key Takeaways

These are the key takeaways from this chapter:

Why care about data leakage?

Data leakage generates subpar predictive performance when the model is deployed in production, relative to the performance you would expect from your test sample. It creates organizational frustration, and you may even put at risk any stakeholder buy-in you may have.

Identifying leakage.

A typical symptom of leakage is having *unusually high* predictive performance on your test sample. You must rely on your knowledge of the problem and the company's experience with these models. It's always a good practice to present your results to more experienced data scientists, and also discuss them with your business stakeholders. If you suspect there's data leakage, you must start auditing your model.

Things to check if you suspect you have data leakage.

Check if you have enforced a strict time windowing process that guarantees that you always predict the future with the past, and not the other way around. Also, check if you have any data transformations where moments or metadata might be leaking.

In ML, scoring reigns.
The litmus test for an ML model is its performance in production. You should direct all of your time and effort to ensure that this is the case.

Further Reading

In my opinion there isn't much depth in most of the published accounts on data leakage found in published books (many mention it just in passing). You can find several useful blog posts on the web: for instance, Christopher Hefele's comment on data leakage at the ICML 2013 Whale Challenge (*https://oreil.ly/j7B4l*) or Prerna Singh's post "Data Leakage in Machine Learning: How It Can Be Detected and Minimize the Risk" (*https://oreil.ly/G92H-*).

Kaufman et al., "Leakage in Data Mining: Formulation, Detection, and Avoidance," (*ACM Transactions on Knowledge Discovery from Data* 6 no. 4, 2012), is a must-read for anyone interested in understanding leakage. They categorize two types of data leakage as those coming from features and those coming from training examples. I decided to deviate a bit from this categorization.

On the problem of complete and quasi-complete separation, the classical reference is A. Albert and J. A. Anderson, "On the Existence of Maximum Likelihood Estimates in Logistic Regression Models" (*Biometrika* 71 no. 1, 1984). A textbook presentation can be found in Chapter 11 of Russell Davison and James MacKinnon, *Econometric Theory and Methods* (Oxford University Press).

The problem of bad controls is well known in the literature of causal inference and causal machine learning. To the best of my knowledge, it was first labeled liked that by Angrist and Pischke in *Mostly Harmless Econometrics* (Princeton University Press). A more recent and systematic study can be found in Carlos Cinelli et al., "A Crash Course in Good and Bad Controls" (*Sociological Methods and Research*, 2022). In this, chapter I used a rather loose version of the bad control definition.

## CHAPTER 12

Productionizing Models

As argued in Chapter 11, the scoring stage reigns in machine learning (ML) since this is the part where all value is created. It's so important that new specialized roles—such as ML engineer and MLOps—have been created to take care of all of the intricacies involved. However, many companies still lack specialized talent, and the job ends up being part of the data scientists' responsibilities.

This chapter provides a helicopter view of production-ready models specifically targeted at data scientists. At the end of the chapter, I will provide some references that will take you deeper into this relatively new topic.

What Does "Production Ready" Mean?

In her book *Designing Machine Learning Systems: An Iterative Process for Production-Ready Applications* (O'Reilly), Chip Huyen states that the process of *productionizing* or *operationalizing* ML entails "deploying, monitoring, and maintaining (a model)." Thus, a working definition for a productionized model is that it has been deployed, monitored, and maintained.

A more direct definition is that a model is *production ready* when it's set for consumption by the end user, be it a human or a system. By *consumption* I mean making use of the predictive scores, which can take place offline or online, and can be done by a human or by another system or service (Figure 12-1).

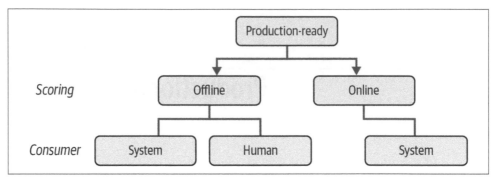

Figure 12-1. Categories of production-ready models

Batch Scores (Offline)

Typically, *batch scoring* entails creating a prediction for a group of rows in a table (be it customers, users, products, or any other such entity) given a set of columns or features. These scores are saved in a table for later consumption.

Batch scoring is very common when:

- The predictive performance isn't greatly improved by having the most up-to-date information.
- You don't have to make a decision with the most up-to-date information.
- You don't have the engineering, infrastructure, or human talent to deploy the model for real-time consumption.

For instance, if you want to predict customer churn over the next month, having granular details of their interactions over the last minute should not improve the quality of the predictions substantially, making batch scoring a suitable method to productionize the model.

Table 12-1 shows an example of how these scores can be saved in a table. Note that the granularity for the table is `customer_id x timestamp` so that you are effectively saving the history of all predictions across customers.

Table 12-1. Example of a table with batch scores

Customer_id	Score	Timestamp
1	0.72	*2022-10-01*
1	0.79	*2022-11-01*
2	0.28	*2022-10-01*
2	0.22	*2022-11-01*
...	...	...

This design might work well for human consumption, as a simple SQL query on an analytical database can be used to retrieve the data; moreover, if you make it part of your data model (say, of your data warehouse), it can be used to create more complex queries that may be needed. Figure 12-2 shows a simplified example of how this can be accomplished. It shows two fact tables (*https://oreil.ly/k05Co*), one that includes the scores from one specific ML model and another one from the business (such as *sales*), and several dimension tables (*https://oreil.ly/5e3uH*). Links between fact and dimensional tables denote that the tables can be joined using primary or secondary keys. Designing the scoring layer as part of your data warehouse can facilitate its consumption, since it allows for easy joining and filtering.

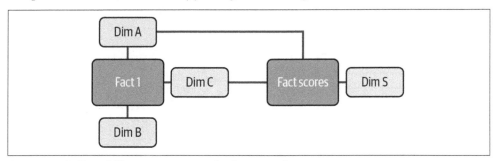

Figure 12-2. Your ML scores as part of a data warehouse

This last design can also work in the case of consumption by a system or service when latency is not one of the main considerations. A typical use case is when scores trigger a communication with your customers (for example, a retention or cross-selling campaign). The pipeline will first query the database, possibly filtering for the top most recent scores, and these customer IDs are then sent to a communication application that sends the emails or SMS (Figure 12-3).

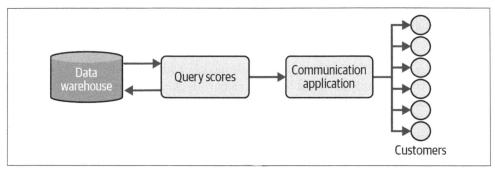

Figure 12-3. Pipeline for system consumption of scores

Real-Time Model Objects

Real-time models are usually not stored as tables, but rather as serialized objects that can be consumed online as new data arrives. Figure 12-4 shows how this process works: your models live in a *model store* that could be an S3 bucket or a more specialized tool such as MLflow (*https://mlflow.org*) or AWS SageMaker (*https://oreil.ly/yzExy*).

The important thing is that these objects can be consumed by another service that takes the most recent data for one specific example (such as a customer or a transaction) to create a single prediction score. As the diagram shows, often the feature vector for one example includes both real-time and batch data. *Importantly, the vector has to match exactly what you used when training the model.*

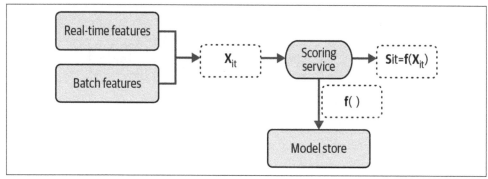

Figure 12-4. Example of online scoring

From this diagram you can already see the complexities that arise in online scoring:

Data architecture
> Your data model should allow for querying of real-time and batch data, so you might end up needing something like a lambda or kappa (*https://oreil.ly/BTYlN*) architecture.

Function as a Service (FaaS)
> Your design should also be able to consume data and the model object *on the fly,* which is commonly achieved through FaaS with cloud computing providers and a microservice architecture. Once you create a score, this is most likely consumed by another service that may, for instance, make a decision, given the score and business rules.

Data and Model Drift

One way of thinking about ML is that you are trying to learn the data generating process (DGP) for an outcome, given some data. When this is done correctly, you can make predictions from similar data:

$$\text{True DGP} : y = f(W)$$

$$\text{Learning the DGP} : \{y, X\}_t \Longrightarrow \hat{f}()$$

The first equation denotes the true DGP that links the outcome variable to a set of true underlying covariates (W). The second equation shows the process of learning this DGP using the data available at a given point in time, that includes the outcome (y) and features (X). Note that the set of features need not coincide with the true underlying covariates.

Since scoring reigns, you should really care about the quality of the predictions *whenever you make those predictions*. The performance of a model can change in time for two main reasons: data drift or model drift. There is *data drift* when the joint distribution for your data changes in time. There is *model drift* when the underlying DGP changes. If you don't retrain your model periodically, data or model drift will generate a decay in the predictive performance. Therefore, you should ensure that you include proper monitoring in your production pipeline, as well as periodic retraining.

Many people have trouble understanding model drift at first, so let me explain the concept with two examples. Suppose you want to attempt a version of Galileo's leaning tower experiment (*https://oreil.ly/k0Apk*) where you let go a resting tennis ball from a chosen height and the objective is to measure the time it takes to hit the ground. You collect measurements of height and time $\{x_t, t\}$, and estimate a linear regression like this:

$$x_t = \alpha_0 + \alpha_1 t + \alpha_2 t^2 + \epsilon$$

The true DGP is given by the laws of physics, and specifically on the surface force of gravity, so it will vary if you run the experiment on Mars or on Earth.[1]

Another example, one closer to the business, has to do with trends and influencers. At the risk of oversimplifying, let me pose that the probability of your product being purchased by some customer i depends on its price and other *stuff*:

$$\text{Prob}\big(\text{purchase}_i = \text{True}\big) = g_i(p, \text{stuff})$$

This is the DGP for customer i, which I don't know, but I'm willing to bet that it might change if all of a sudden Jungkook (*https://oreil.ly/oFkaY*) starts promoting it on social media. Specifically, I would expect a lower price sensitivity for customers in

---

1 Recall that with constant velocity and acceleration $x = x_0 + v_0 t + \frac{g}{2}t^2$, where g is the surface force of gravity. On Mars (*https://oreil.ly/LmYrT*), the force of gravity is ~38% of Earth's.

the segment who like Korean pop and follow him. The old DGP $g_i()$ has drifted to something like $\tilde{g}_i()$.

A Cautionary Tale of Model Drift: Zillow Offers

In 2021, the real estate marketplace Zillow suffered a loss of more than $500 million in one of the best known cases of model drift. To understand what happened, imagine that you have a good prediction model for property prices; you can then buy properties if you predict that their price will increase, and make a profit on the price differential (buy cheap, sell expensive).

Zillow Offers was a product that attempted to do exactly that. Long story short, at the beginning it did great, so the company scaled the product, but at some point the model started drifting and the predictions were no longer good. The company ended up owning properties that could only be sold at a loss. Had Zillow monitored and retrained its model, it would've had a chance to learn the new DGP and make the correct buying decisions.

Essential Steps in any Production Pipeline

Figure 12-5 shows the minimal required steps that most ML pipelines should include. Following the recommendation in Chapter 11, I have separate tasks for the scoring and training stages, but they share several stages. I also use a lighter shade of gray to denote stages where you store metadata for monitoring purposes. I will now describe each stage in more detail.

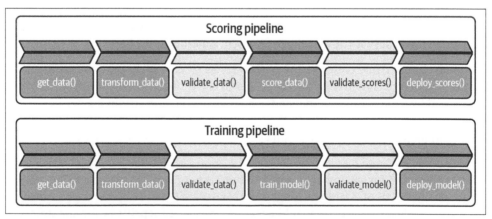

Figure 12-5. Generic production pipelines

Get and Transform Data

As the names suggest, the `get_data()` stage creates a connection and queries the data source; given this raw data, the `transform_data()` step applies a predefined set of in-memory transformations on some or all of the columns of your table. It's not uncommon for the first method to be SQL-based, while the latter can be run using Python (or Spark).

While I have separated the two stages, depending on the nature of the problem it may be advisable to merge these into one unique stage. Let's consider the pros and cons of establishing this modular separation.

Although modularization is usually a good practice—allowing for cleaner and more expedited debugging and data and model governance—it can impose computational costs or restrictions that can be better addressed by *pushing* both operations to the querying engine. This is especially true since querying engines are normally optimized and resourced to handle large datasets, while you may end up with fewer resources to transform the smaller subset of the data actually needed.

On the other hand, SQL is great to query tabular data, but it may not provide enough flexibility to create complex transformations that are easier to achieve with Python or Spark.

Moreover, separation allows for a fully focused and independent transformation stage. This is important because feature engineering (*https://oreil.ly/R6dhb*) plays a critical role in developing performant ML models. Hence, modularization allows for a more thorough documenting and reviewing of the main transformations of your model.

Finally, breaking each stage into independent modules is great to allow more expedited code reviewing and thus shorter deployment cycles.

> If you are memory constrained to make some transformations, but your querying engine can perform some high-memory computations, it's sometimes advisable to *push* some or all of the transformations to the querying stage.

You may rightly suspect that these stages are shared by the training and scoring pipelines, explaining why I decided to use a function-like notation. If you use a windowing methodology, such as the one described in Chapter 11, the `get_data()` method can easily be parameterized to query data for a given time window.

Outputs for the `transform_data()` stage in the training pipeline are the final arrays needed to train your model; for supervised learning, it would be something like this:

```
transform_data(get_data(Data)) ⇒ y, X
```

For the scoring data, it would only be an array of features X.

Validate Data

This is the first monitoring stage of each pipeline, and is used to store metadata and alert for the presence of data drift. This stage can be decomposed into two substages:

1. *Calculate and save statistics.* Compute a set of predefined statistics for the distributions of the output and features, and save in a table.

2. *Test for data drift.* Given the current and historical statistics just saved, run a test to see if the changes are pure noise or signal. The output should be either creating an alert or not.

Table 12-2 shows an example of a table that stores distribution metadata for all of a company's models. This table can be used to store deciles for outcome and features for all models, and for the training and score stages, so it's easy to use for reporting, testing, and monitoring just by applying filters.

Table 12-2. Example of a table with deciles

Model	Stage	Metric	Decile	Value	Timestamp
churn	training	outcome	d1	100	2022-10-01
...	...	outcome	...	...	...
...	...	outcome	d10	1850	2022-10-01
churn	training	feature1	d1	-0.5	2022-10-01
...	...	feature1	...	...	...
...	...	feature1	d10	1.9	2022-10-01
...	...	...	...	...	...

In this example, I chose to save the deciles for each variable in the dataset, since these capture quite a bit of information from the corresponding distributions.

For testing, there are many alternatives. If you have enough history and want to follow a traditional hypothesis testing route, you can run regressions for each metric and decile ($d_{m,t}$), such as this:

$$d_{m,t} = \alpha_m + \beta_m t + \epsilon_{m,t}$$

where t, as a feature, denotes a time trend: if the p value for β_m is lower than a desired threshold level (10%, 5%, 1%), you can reject the null that the parameter is 0, so you have evidence of metric m drifting.

Alternatively, you can use a nonparametric test similar to the ones used in Chapter 9 where you calculate upper and lower quantiles in the historical distributions and check whether the new observation lies within that confidence interval (for example, to compute 95% confidence intervals, you calculate $q_{2.5\%}$, $q_{97.5\%}$).

Some people prefer to run Kolmogorov-Smirnov (*https://oreil.ly/4j73f*) tests, so you might need to save a different set of metadata, but the logic is the same.

 Whatever you decide to use, my recommendation is to *keep it simple*. Often, all you need is a dashboard that plots this metadata, which enables you to set up simple alerts for when changes occur.

When you productionize a model, it's often the case that the simpler, the better.

Training and Scoring Stages

Once you have your training data ready, you can proceed with the formal process of training where you usually do the following:

1. Divide the sample into training, test, and validation subsamples.
2. Optimize metaparameters and minimize a loss function.

The output from the `train_model()` stage is a model object that can be used for prediction purposes:

```
train_model(transform_data(get_data(Data))) ⇒ f()
```

Similarly, the `score_data()` method uses some features X to create a prediction or score:

```
score_data(transform_data(get_data(Data)), f()) ⇒ ŝ
```

As mentioned earlier, this score can be saved on a table for offline consumption or passed to another service for online consumption.

Validate Model and Scores

Before moving on, it's a good practice to save some metadata again that will help create alerts for model or data drift. In this stage, I like to create the same metadata in `validate_data()`, but only passing the test sample scores (`validate_model()`) or the actual scores (`validate_scores()`). If you follow this route, you actually *reuse* the previous method, but just pass a different dataset across stages and pipelines; everything else is taken care of (such as updating the metadata table and alerting).

Note that for online consumption you need to gather enough data for validation, but the logic is essentially the same.

Deploy Model and Scores

As the names suggest, the objective of these stages is to save the model and scores. For the training pipeline, you will have to serialize the model object and save it using some persisting storage (such as disk, S3 bucket, or the like). Adopting good naming (*https://oreil.ly/r8tzX*) and versioning (*https://semver.org*) conventions will help with cataloging the models.

The topic of model serialization is important and technical, so I will provide more references at the end of this chapter.

The deployment of the scores depends on whether consumption is offline or online. In offline scoring, you just write the scores in a table to make it available for consumption. In online scoring, not only should you make the score available for consumption by another service, but you should also store it in a table.

Key Takeaways

These are the key takeaways from this chapter:

Scoring reigns. Productionizing your model should be at the top of your priorities since only productive models can create value for an organization.

What is production ready?
   A model is productive when it is ready for consumption. Since most of the time a model will be consumed at different periods of time, you must create a process to guarantee that the model has enduring predictive performance.

Model and data drift.
   There is model drift when the data generating process for your outcome changes. Data drift refers to changes in the distribution of your outcome or features. When left unhandled, data and model drift will create a decay in your model's predictive performance over time. The best way to avoid drift is to retrain your models in a recurrent way.

Production pipelines.
   It's good to set a minimal structure for your production pipelines. Here I propose to have modular and separate training and scoring pipelines that share some of the methods or stages. Critically, you should include stages where you create and store metadata that will alert you if there's model or data drift.

Keep it simple.

Deploying in production is a complex sequence of steps, so the recommendation is to keep each of these as simple as possible. Unnecessary complexity may end up compounding, making it very hard to find the source of a problem when this comes up (and it will come up).

Further Reading

Written by an industry expert, Chip Huyen's *Designing Machine Learning Systems* is great, providing many of the critical technical details left out in this chapter. I cannot recommend this enough.

I found Valliappa Lakshmanan et al., *Machine Learning Design Patterns: Solutions to Common Challenges in Data Preparation, Model Building, and MLOps* (O'Reilly), very useful. The aim is to synthesize a set of ML design practices that can be used across the board. Since it was written by three Google engineers, you will find that their examples rely extensively on Google's infrastructure, so many times it's not obvious how to translate that to other cloud service providers. But if you're able to abstract away this nuisance, you'll find this book a great read and resource.

Kurtis Pykes's blog post "5 Different Ways to Save Your Machine Learning Model," (*https://oreil.ly/2Lsuq*) discusses different ways to serialize your ML model.

Lu et al., "Learning under Concept Drift: A Review," (April 2020, retrieved from arXiv (*https://oreil.ly/3dRLZ*)), present a comprehensive review of *concept drift*, which is sometimes (*https://oreil.ly/RBHY2*) taken to encompass both data and model drift.

On the Zillow Offers model drift case, you can read the *MarketWatch* article by Jon Swartz (November 2021), "Zillow to Stop Flipping Homes for Good as It Stands to Lose More Than $550 Million, Will Lay Off a Quarter of Staff," (*https://oreil.ly/J-lWA*) or Anupam Datta's "The Dangers of AI Model Drift: Lessons to Be Learned from the Case of Zillow Offers" (*https://oreil.ly/NMo5A*) (*The AI Journal*, December 2021).

## CHAPTER 13

Storytelling in Machine Learning

In Chapter 7, I argued that data scientists ought to become better storytellers. This holds true in general, but it takes on special importance with regard to machine learning (ML).

This chapter walks you through the main aspects of storytelling in ML, starting with feature engineering and finishing with the problem of interpretability.

A Holistic View of Storytelling in ML

Storytelling plays two related but distinct roles in ML (Figure 13-1). The better-known role is a salesperson, where you need to engage with an audience, possibly to gain or maintain stakeholder buy-in, a process that usually takes place after you've developed a model. The lesser-known role is a scientist, where you need to find hypotheses that will guide you throughout the process of developing the model.

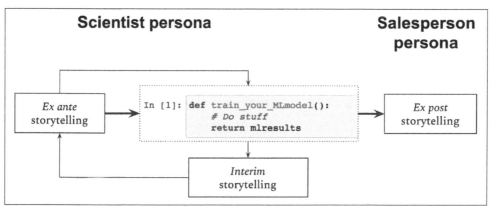

Figure 13-1. Storytelling in ML

Since the former takes place *after* you have developed your model, I call it *ex post* storytelling; your scientist persona is mostly invoked before (*ex ante*) and during (*interim*) the process of training the model.

Ex Ante and Interim Storytelling

Ex ante storytelling has four main steps: defining the problem, creating hypotheses, feature engineering, and training the model (Figure 13-2). While they usually flow in that direction, there's a feedback loop between all of them, so it's not uncommon that after you train a first model, you iterate on the features, hypotheses, or even on the problem itself.

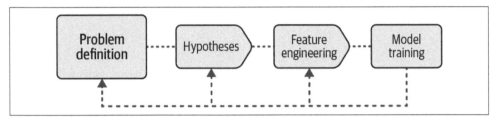

Figure 13-2. Ex ante storytelling

The first step is always the problem definition: *what* do you want to predict and *why*? This is better done early and collaboratively with your stakeholders to ensure you have their buy-in, as many promising ML projects fail because of this.

Recall from Chapter 12 that a model is only good if it has been deployed in production. Deploying to production is a costly endeavor, not only in terms of time and effort, but also in terms of any alternative project you could've been working on (opportunity cost). Because of this, it's always good to ask yourself: *do I really need an ML implementation for this project?* Don't fall into the trap of doing ML just because it's sexy or fun: your objective should always be to create the maximum value, and ML is just one more tool in the bag.

Finally, in the problem definition, don't forget to have good answers to the questions:

- How is this model going to be used?
- What are the levers that can be pulled using predictions from the model?
- How does it improve your company's decision-making capabilities?

Having sound answers to these questions will help the business case for developing an ML model, thereby increasing the likelihood of success.

As a general recommendation, the sooner you involve your stakeholders in the definition of the problem, the better. This helps with having stakeholder buy-in from the outset. Also ensure that ML is the appropriate tool for the problem at hand: deploying, monitoring, and maintaining a model are costly, so you should have a good business case for it.

Creating Hypotheses

With a well-defined problem, you can now switch into your scientist persona and start creating hypotheses for the problem at hand. Each of these hypotheses is a story about the drivers for your prediction; it's in this specific sense that scientists are also storytellers. Successful stories improve the predictive performance of your model.

At this point, the key questions are: *what am I predicting, and what drives this prediction*? Figure 13-3 shows a high-level overview of the types of prediction problems and their relationship to the levers at your disposal. Understanding the levers is critical to ensure that an ML model creates value (Chapter 1).

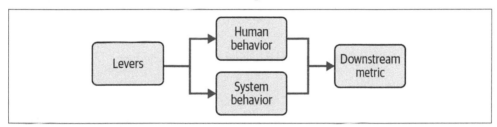

Figure 13-3. Lever-behavior-metrics flow

From here it follows that most prediction problems fall into one of these categories:

Metrics that arise from human behavior
> Many times, the metric that you care about depends on your customers acting in some specific way. For instance, will my user click on a banner? Will they purchase the product at the reference price? Will they churn next month? How much time will they spend on the marketplace?

Metrics that arise from systems behavior
> Metrics depend also on how your systems perform. One of the best well-known examples is data center optimization, and most specifically, cracking the air cooling problem (*https://oreil.ly/5guWh*). Another is predicting the loading time for your web page, which has been found (*https://oreil.ly/xXtbS*) to directly impact churn metrics.

Downstream metrics

Many times you just care about aggregate downstream metrics, such as revenue. This is most common with data scientists working directly in financial planning and analysis (FP&A).

 Many data scientists struggle with the process of creating and engineering features that are predictive. A general recommendation is to always start by writing down and discussing with others a list of hypotheses for the prediction problem. Only then should you move forward with the process of feature engineering. Don't forget to write down the reasons you believe a hypothesis might be right. Only with this rationale will you be able to challenge your logic and improve upon a given story.

Some high-level recommendations to come up with hypotheses for your problem are:

Know your problem really well.

The not-so-secret sauce to building great ML models is to have substantial domain expertise.

Be curious.

This is one defining trait that makes a data scientist a scientist.

Challenge the status quo.

Don't be afraid to challenge the status quo. This includes challenging your own hypotheses and iterating when needed (be aware of any signs of confirmation bias on your side).

This said, let's go into some more specific recommendations on how to proceed on your hypothesis discovery and formulation.

Predicting human behavior

For predicting human behavior, it's useful to always remember that people do what they *want* and *can* do. You may want to go to Italy, but if you can't afford it (money or timewise), you won't do it. Tastes and resource availability are of first-order importance whenever you want to predict human behavior, and this can take you a long way toward coming up with hypotheses for your problem.

Thinking about motivations will also force you to think really hard about your product. For instance, why would anyone want to buy it? What is the value proposition? Which customers would be willing to pay for it?

Challenge the Status Quo: Lessons from the Trench

Not so long ago, a data scientist from one of my teams was working on a prediction model to cross-sell a relatively new product that was having trouble gaining traction and scale. This was a product with a relatively weak value proposition, so understanding which customers would be willing to use and pay for it was very hard (and thus so was building a model to predict that!).

I worked with her and after doing some hard work toward understanding the product, the value proposition, and our customers, we came back with the recommendation that unless there was some serious redesigning of the product, we would not have product-market fit. It took us almost one year to convince the stakeholders that this was indeed the case, and at certain points several of them were not happy with us.

Another trick is to use your capacity to empathize with your customers; ask yourself what would *you* do if you were them? Of course, the easier it is to put yourself in their shoes, the better (for me it would be really hard to put myself in an influencer's or professional boxer's shoes). This trick can take you far, but bear in mind that you may not be your typical customer, which brings me to the next trick.

At least at the beginning, aim for understanding and modeling your *average* customer. You should first and foremost get first-order effects right, meaning that modeling the average unit of analysis will buy you quite a bit of predictive performance. I've seen many data scientists start hypothesizing about corner or edge cases which, by definition, will have a negligible impact on overall predictive performance. Corner cases are interesting and important, but for prediction, it's almost always better to start with the average cases.

Predicting system behavior

Some of the previous remarks also apply for predicting a system. The main difference is that since systems lack purpose or sentience, you can restrict yourself to understanding technical bottlenecks.

Clearly, you have to master the technical details of your system, and the more knowledgeable you become about the physical constraints, the easier it will be to come up with hypotheses.

Predicting downstream metrics

Downstream metrics prediction is both harder and easier than predicting individual metrics that result from human or system behavior. It's harder because the more distanced from the underlying drivers the metric is, the weaker and more diffused your hypotheses become. Moreover, it inherits the difficulty of coming up with stories about these drivers, and some of these may compound and create higher-level complexity.

This said, many times you can do some hand-waving and exploit the time and space correlations to create some features. In a sense, you're accepting that any stories you come up with will be beaten by a simple autoregressive structure that is common in time series and spatial autoregressive models.

Feature Engineering

Generally speaking, the process of feature engineering entails converting hypotheses into measurable variables that have enough signal to help your algorithm learn the data generating process. It's a good practice to split this into several stages, as depicted in Figure 13-4.

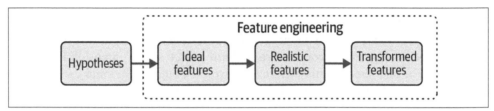

Figure 13-4. Feature engineering flow

The stages for feature engineering are:

Create a set of ideal features.
> The first step is about translating your hypotheses into *ideal* features, if you were able to measure everything precisely. This step is important, as it allows you to set a baseline for the second stage.

> An example is the role that *intentionality* has on *early churn*, defined as those customers that try a product once and leave. One hypothesis is that these customers didn't really intend to use the product (because they were just trying it, or the sale was pushed (*https://oreil.ly/HDGj-*), or there was sales fraud, or the like). Wouldn't it be great if you could ask them and they answered truthfully? Unfortunately, this isn't practical or achievable.

Approximate the ideal features with realistic features.
> If you realize that the ideal set of features is unavailable, you need to find good proxy features, that is, features that are correlated with the ideal ones. Many times, the degree of correlation can be very low, and you need to settle for including controls with a very weak correspondence to the original hypothesis.

> An example of the latter is how culture affects your tastes and thus your likelihood to purchase a product. For instance, there may be cultural differences to explain why users in different countries decide to accept or reject the cookies in

their browser (people from some countries may be more sensitive to sharing this information). Needless to say, measuring culture is hard. But if you suspect that country-level variation will capture a big part of the variation of the cultural hypothesis, all you need is to include country dummy variables. It's a relatively weak set of features because these will proxy any feature at this level, and not only culture (for instance, differences in regulatory environments).

Transform features.

This is the process of extracting the maximal amount of signal from your features by applying a set of transformations on them. Note that I'm departing a bit from the literature, since most textbook treatments on feature engineering refer exclusively to this stage.

This stage involves transformations such as scaling (*https://oreil.ly/Hak0v*), binarizing and one-hot encoding (*https://oreil.ly/ralbT*), imputation of missing values (*https://oreil.ly/MhGuK*), feature interactions (*https://oreil.ly/bT-1q*), and the like. I provide several references at the end of this chapter where you can consult the vast array of available transformations.

Importantly, transformations depend on your data *and* the algorithm of your choice. For instance, with classification and regression trees you may not need to take care of outliers by yourself, since the algorithm will do it for you. Similarly, with generally nonlinear algorithms, like trees and tree-based ensembles, you need not include multiplicative interactions.

Example: Predicting Sales

Suppose that you want to predict sales at a geographical area (g). A typical use case for such a model is when you want to direct your sales force to locations with the highest sales potential, as predicted by the model.

I'll use a trick from Chapter 2 to get crisper stories:

$$\text{sales}_g = \text{TAM}_g \times \frac{\text{sales}_g}{\text{TAM}_g} = \text{TAM}_g \times \text{Prob}(\text{unit sale in } g)$$

This just says that total sales in unit g must equal the total addressable market (TAM), multiplied by the probability of a sale in that area.

By doing this, instead of coming up with hypotheses for the number of sales per location, I can now focus on stories that will help me predict TAM *and* stories to explain why the company makes a sale. The latter involves human behavior, and the former is an aggregate metric.

To model TAM, I need to first understand who is my target customer and then find stories about what makes them cluster in certain locations. For instance, to predict the TAM for this book, I want to estimate the number of data scientists in a given location. One plausible story is that data scientists are where companies need them. I can further refine this story by arguing that company size matters (*because* of the amount of data needed to make the business case for the data scientist positive, but also because data scientists are relatively expensive and only large enough companies can hire them), that industry mix matters (*because* more capital-intensive industries may have more automated systems-generated data than more labor-intensive industries with more manual processes, or *because* of regulatory pressures, or *because* of market concentration differences), and that population size and age distributions matter (*because* the field is relatively new and younger people, but not too young, are more willing to invest in learning a hard technical subject like data science). These hypotheses guide what type of data I need to look for to solve this prediction problem.

To model the probability of a sale being made, there must be people who want and can afford the product (demand), and the product must be available to them in those locations (supply). Ideal features to model demand are consumer preferences for the product as well as household income. Preferences are generally hard to get, but can be approximated with the company's previous sales per location, or with online search behavior (such as Google trends or data available by similar vendors). Supply-side data is easier to get since I should know if the company and its competitors have a presence in different locations.

Ex Post Storytelling: Opening the Black Box

The problem of ex post storytelling is mainly one of understanding why your model makes predictions as it does, what are the most predictive features, and how these are correlated to predictions. The two main points you want to convey to your audience are:

- The model is incrementally predictive, that is, the prediction error is lower than that of the baseline alternative.

- The model *makes sense*. A good practice is to start discussing the hypotheses, how they were modeled, and how they are consistent with the results.

Generally speaking, a model is *interpretable* if you can understand what drives its predictions. *Local* interpretability aims at understanding specific predictions, such as why a customer is deemed highly likely to default on a credit. *Global* interpretability aims at providing a general understanding of how features affect the outcome. This topic deserves a book-length presentation, but in this chapter I can only delve into the more practical matters, and specifically, I will only go through methods to achieve global interpretability, as I've found these to be most useful for storytelling purposes.

Before opening the black box, be sure that your model has enough predictive performance, and that there's no data leakage. You'll need to devote enough time and effort into ex post storytelling, so you'd better start with a good prediction model.

Also, when presenting performance metrics, try to make them as relatable to your audience as possible. Common metrics, such as the root mean square error (RMSE) or the area under the curve (AUC), can be cryptic for your business stakeholder. It's generally worth the effort to translate them to precise business outcomes. For instance, if you have a 5% lower RMSE, how is the business better?

Interpretability-Performance Trade-Off

It can be argued that an ideal ML algorithm is both performant and interpretable. Unfortunately, there is usually a trade-off between interpretability and predictive performance, so you have to give up part of your understanding of what's happening inside the algorithm if you want to achieve lower prediction error (Figure 13-5).

On one side of the spectrum, you have linear models that are generally considered to be highly interpretable but have subpar predictive performance. This set includes linear and logistic regression, as well as nonlinear learning algorithms, such as classification and regression trees. On the other side of the spectrum are the more flexible, and usually highly nonlinear, models, like deep neural networks, tree-based ensembles, and support vector machines. These algorithms are generally known as *black box* learners. The objective is to open the black box and gain a better understanding of what's going on.

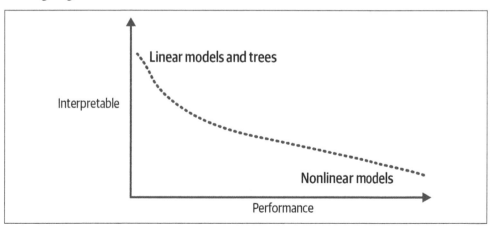

Figure 13-5. Interpretability-performance trade-off

Before moving on, it's not obvious that you need to interpret the results, so let's briefly discuss why you may want to do so:

Adoption and buy-in
> Many people need to understand why a prediction is made in order to accept it as valid, thereby adopting it. This is most common in organizations that are not used to the ML approach, and decisions are usually made using a quasi-data-driven approach that involves a lot of gut instinct. You may find it easier for your stakeholders to accept your results and sponsor your project if you are able to open the black box for them.

Low real-world predictive performance
> Opening the black box is one of the most effective ways to detect and correct problems like data leakage (Chapter 11).

Ethics and regulatory requirements
> In certain industries it's actually required that companies explain why a certain prediction was made. For instance, in the US (*https://oreil.ly/5zj9j*), the Equal Opportunity Act entitles anyone to ask for the reasons why a credit was denied. A similar criterion applies with the European General Data Protection Regulation (GDPR). Even if you are not required to, you may want to validate whether the predictions and subsequent decisions follow a minimal ethical standard by opening the black box.

Linear Regression: Setting a Benchmark

Linear regression provides a useful benchmark to understand interpretability (see also Chapter 10). Consider the following simple model:

$$y = \alpha_0 + \alpha_1 x_1 + \alpha_2 x_2 + \epsilon$$

By making strong linearity assumptions about the underlying data generating process, you immediately get:

Effect directionality
> The sign of each coefficient tells you if the feature is positively or negatively correlated with the outcome, after controlling for all other features.

Effect magnitude
> Each coefficient is interpreted as the change in the outcome associated with a one-unit change in each feature, holding other features fixed. Importantly, no causal interpretation can be given without further assumptions.

Local interpretability
> From the first two items, you can assert why any individual prediction was made.

Some data scientists make the mistake of giving the absolute magnitude of the coefficients a relative *importance* interpretation. To see why this doesn't work, take the following model, where revenue is expressed as a function of the size of the sales force and paid marketing spend (search engine marketing or SEM):

$$\text{revenue} = 100 + 1000 \times \text{Num. sales execs} + 0.5 \times \text{SEM spend}$$

This says that, on average and holding other factors fixed, each additional:

- Sales executive is associated with an increase of $1,000 in revenue.
- Each dollar spent on SEM (for example, bids on Google, Bing, or Facebook ads) is associated with an increase of 50 cents in revenue.

You would be tempted to conclude that increasing the size of the sales force is *more important* for your revenues, compared to paid marketing spend. Unfortunately, this is an apples-to-oranges comparison since each feature is measured in different units. A trick to measure everything in the same units is to run a regression on standardized features:

$$y = \beta_0 + \beta_1 \tilde{x}_1 + \beta_2 \tilde{x}_2 + \eta$$

$$\text{where} \quad \tilde{z} = \frac{z - mean(z)}{std(z)} \quad \text{for any variable } z$$

Note that regression coefficients on standardized variables are generally different from those in the original model (hence the different greek letters), and thus have a different interpretation: by standardizing all features, you measure everything in units of standard deviations (*unitless* is a better term), ensuring that you compare apples to apples. You can then say things like: x_1 *is more important than* x_2, *since an additional standard deviation in* x_1 *increases revenue by more than a corresponding increase in* x_2.

The trick here is to find a way to convert the original units into a common unit (in this case, standard deviations). But any other common unit could also work. For instance, imagine that each additional sales executive costs $5,000 per month, on average. Since marketing spend is already in dollars, you end up saying that on average, each additional dollar spent in:

- Sales executives is associated with a 20 cent increase in revenue
- Paid marketing is associated with a 50 cent increase in revenue

While this last method also works, standardization is a much more common method to find a common unit for all features. The important thing to remember is that you're now able to *rank* features in some meaningful way.

Figure 13-6 plots the estimated coefficients, along with 95% confidence intervals, for a simulated linear model with two zero-mean, normally distributed features (x_1, x_2), as in the previous equations. Features z_1, z_2, z_3 are additional variables correlated to x_2, but are otherwise unrelated to the outcome. Importantly, I set the true parameters to $\alpha_1 = \alpha_2 = 1$ and $Var(x_1) = 1$, $Var(x_2) = 5$. This has two effects:

- It increases the signal-to-noise ratio for the second feature, thereby making it more informative.
- It increases the true coefficient:[1] $\beta_2 = \sqrt{5}\alpha_2$.

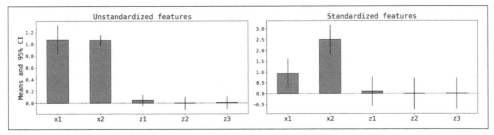

Figure 13-6. Regression in linear versus standardized features

By standardizing both features, it becomes noteworthy that the second feature ranks higher in terms of importance, as defined earlier. Thanks to the confidence intervals, you can also conclude that the last three features are uninformative. An alternative to the statistical approach would be to use *regularization*, such as in a Lasso regression.

Feature Importance

Many times you want to rank the features according to some objective measure of importance. This is useful for ex ante and ex post storytelling purposes. From an ex post point of view, you can say things like: *we found that the time of the transaction is the most important predictor of fraud*, which might help you sell the result of your model, and will also deliver potentially great Aha! moments for you and your audience (see also Chapter 7).

From an ex ante point of view, having a way to rank features by importance can help you iterate on your hypotheses or feature engineering, or improve your understanding of a problem. If you have well-thought-out hypotheses and your results look suspicious, it's more likely that you made a programming error on the feature engineering side, or that you have data leakage.

---

[1] It's easy to show that in linear regression, rescaling a feature x to kx changes the true coefficient from α to α/k.

Earlier, I used standardized features in a linear regression to get one possible such importance ranking:

Standardized feature importance in linear regression
A feature x is more important than feature z if a one standard deviation increase in x is associated with a larger change in the outcome, in absolute value.

Alternatively, *importance* can be defined in terms of each feature's amount of information content for the prediction problem at hand. Intuitively, the higher the information content of a feature (for a given outcome), the lower the prediction error if the feature is included. There are two commonly used metrics that follow this route:

Impurity-based feature importance
A feature x is more important than feature z, from a node impurity point of view, if the relative improvement in prediction error from nodes where x was chosen as a splitting variable is larger than the corresponding increase for z.

Permutation importance
A feature x is more important than feature z, from a permutation point of view, if the relative loss in performance when the values of x are permuted is larger than that for z.

Note that impurity-based feature importance (*https://oreil.ly/acJDH*) only works for tree-based ML algorithms. Every time a node is split using a feature, the improvement in performance is saved, so at the end you can compute the share of improvements for all features relative to the total improvement. With ensembles, this is the average across all trees grown.

On the other hand, permutation importance (*https://oreil.ly/84XXY*) works with any ML algorithm since you just shuffle the values of each feature (several times, as in a bootstrapping procedure) and compute the loss in performance. The intuition is that the actual order matters more for *important* features, so there should be a larger loss in performance from the permutation of values.

Figure 13-7 shows permutation and impurity-based feature importances using the same simulated dataset as before, trained with a gradient boosting regression (no metaparameter optimization), along with 95% confidence intervals. Confidence intervals for permutation importances are computed parametrically (assuming normality) using the means and standard deviations provided by scikit-learn. I obtain analogous intervals for impurity-based features using bootstrapping (see Chapter 9).

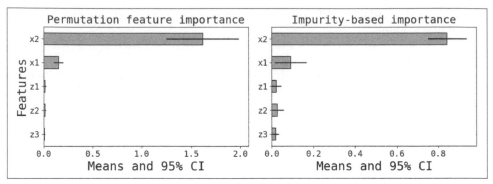

Figure 13-7. Feature importances for simulated model using gradient boosting regression

Heatmaps

Heatmaps are very easy to compute and are generally quite good at visually displaying the correlation between each feature and the predicted outcome. This is quite handy to say things like *when x increases, y falls*. Many hypotheses are stated directionally, so a quick first test of whether this holds in practice is quite useful. The process to calculate them is as follows:

1. Split the predicted outcome (regression) or probability (classification) in deciles, or any other quantile.

2. For each feature x_j and decile d, calculate the average across all units in that bucket: $\bar{x}_{j,d}$.

These can be arranged in a table with deciles in the columns, and features in the rows. It's usually good to order the features using some measure of importance so that you focus on the most relevant features first.

Figure 13-8 shows a heatmap for the linear regression trained on the previous simulated example, where features have already been sorted by feature importance. Just by inspecting the relative shades for each feature (row), you can easily identify any patterns, or lack thereof.

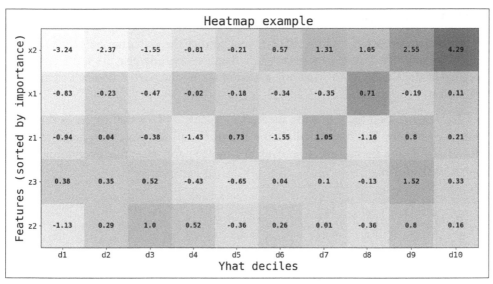

Figure 13-8. Feature heatmap for previous simulated example

For instance, x_2 is positively correlated with the outcome, as expected since the true coefficient in the simulation is equal to one. Units in the lower decile have -3.58 units on average, and this increases monotonically up to 4.23 units on average for the top decile.

Inspecting the row for x_1 shows the main problem that heatmaps have: they present bivariate correlations only. The true correlation is positive ($\alpha_1 = 1$), but the heatmap fails to capture this monotonicity. To understand why, notice that x_1 and x_2 are *negatively* correlated (Figure 13-9). However, the larger variance of the second feature gives it more predictive power, and thus more weight in the final ordering of the predicted outcome (and deciles). These two facts break the monotonicity that was expected for the second feature.

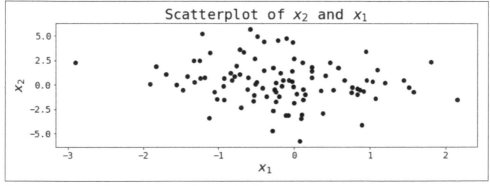

Figure 13-9. x_2 and x_1 are negatively correlated

Partial Dependence Plots

With partial dependence plots (PDPs), you predict the outcome or probability by only changing one feature at a time, while fixing everything else. It's quite appealing because of the similarity to what you get from taking the partial derivatives in linear regression.

In Chapter 9, I used the following method to calculate PDPs that captures this intuition very closely. You first calculate the means for all features, then create a linear grid of size G for the feature you want to simulate, and assemble everything into a matrix of the form:

$$\overline{\mathbf{X}}_{\mathbf{j}} = \begin{pmatrix} \bar{x}_1 & \bar{x}_2 & \cdots & x_{0j} & \cdots & \bar{x}_K \\ \bar{x}_1 & \bar{x}_2 & \cdots & x_{1j} & \cdots & \bar{x}_K \\ \vdots & \vdots & \ddots & \vdots & & \vdots \\ \bar{x}_1 & \bar{x}_2 & \cdots & x_{Gj} & \cdots & \bar{x}_K \end{pmatrix}_{G \times K}$$

You then use this matrix to create a prediction with your trained model:

$$\text{PDP}^{(1)}\left(x_j\right) = \hat{f}\left(\overline{\mathbf{X}}_{\mathbf{j}}\right)$$

This method is fast and intuitively appealing, and it also allows you to quickly simulate the impact of interactions between features. However, from a statistical point of view, it's not really correct since the average of a function is generally different from the function evaluated on the averages of the inputs (unless your model is linear). The main advantage is that it requires only one evaluation of the trained model.

The correct way to do it—and the method used by scikit-learn (*https://oreil.ly/waddK*) to compute PDPs—requires N (sample size) evaluations of the trained model for each value g in the grid. These are then averaged out to get:

$$\text{PDP}^{(2)}\left(x_j = g\right) = \frac{1}{N}\sum_{i=1}^{N} \hat{f}\left(x_{1,i}, \cdots, x_{j-1,i}, g, x_{j+1,i}, \cdots, x_{K,i}\right)$$

Interactions can be easily simulated by changing several features at a time. In practice, often the two methods provide similar results, but this really depends on the distribution of the features and the real unobserved data generating process.

Before moving on, notice that in this last computation you have to compute a prediction for each row in your dataset. With *individual conditional expectation (ICE) plots*, you visually display these effects across units, making it a method of local interpreta-bility, as opposed to PDPs.[2]

Let's simulate a nonlinear model to see the two methods in action, using the following data generating process:

$$y = x_1 + 2x_1^2 - 2x_1x_2 - x_2^2 + \epsilon$$
$$x_1 \sim Gamma(\text{shape} = 1, \text{scale} = 1)$$
$$x_2 \sim N(0, 1)$$
$$\epsilon \sim N(0, 5)$$

I use a gamma distribution for the first feature to highlight the effect that outliers may have when you use either method.

Figure 13-10 shows the estimated and true PDPs using both methods. PDPs for the first feature capture well the shape of the true relationship, but the two methods start diverging from each other for larger values of x_1. This is expected because the sample mean is sensitive to outliers, so with the first method you end up using an average unit with a relatively large first feature. With the second method, this isn't as pro-nounced since individual predictions are averaged out, and in this particular example the functional form smooths out the effect of the outliers.

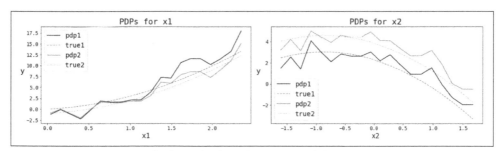

Figure 13-10. PDPs using both methods in the simulated data

While PDPs are great, they are biased with correlated features. For instance, if x_1 and x_2 are positively correlated, both will then have small or large values at the same time. But with a PDP you may end up unrealistically imposing a small value (from the grid) for x_1 when the corresponding value for the second feature is large.

To see this in practice, I simulated this modified version of the previous nonlinear model:

---

2 The implementation on the code repo (*https://oreil.ly/dshp-repo*) provides the ICE and the PDP.

$$y = x_1 + 2x_1^2 - 2x_1x_2 - x_2^2 + \epsilon$$

$$x_1, x_2 \sim N(\mathbf{0}, \Sigma(\rho))$$

$$\epsilon \sim N(0, 5)$$

where the features are now drawn from a multivariate normal distribution with a covariance matrix indexed by a correlation parameter. Figure 13-11 plots the estimated and true PDPs for uncorrelated ($\rho = 0$) and ($\rho = 0.9$) correlated features, where you can readily verify that PDPs are biased when features are correlated.

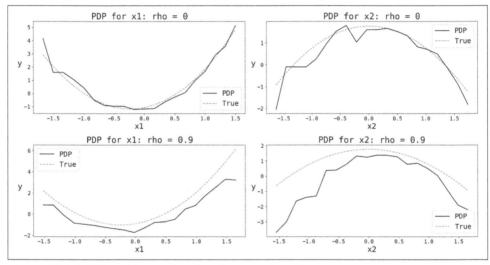

Figure 13-11. PDPs with correlated and uncorrelated features

Accumulated Local Effects

Accumulated local effects (ALE) is a relatively new method that takes care of the shortcomings of PDPs when handling correlated features. It's also less computationally expensive since the number of evaluations of the trained function is smaller.[3]

As discussed, the problem with PDPs arises from imposing unrealistic values of a feature given its correlation with the remaining ones, which end up biasing the estimates. As before, you start by creating a grid for any feature k under inspection. ALE handles this by doing three things:

---

3 At the time of writing, two Python packages are available that calculate ALEs: ALEPython (*https://oreil.ly/znDHe*) and alibi (*https://oreil.ly/QlZkS*). You can find my own implementation for the case of continuous features and no interactions in the code repo (*https://oreil.ly/dshp-repo*).

Focusing on local effects

For a given value in the grid g, select only those units (i) in your data for which the value of the feature falls in a neighborhood of that point ($\{i: g - \delta \leq x_{ik} \leq g + \delta\}$). With correlated features, all of these units should have relatively consistent values for all other variables.

Computing the slope of the function

Within that neighborhood, you compute the slope for each unit, and these are then averaged out.

Accumulating these effects

For visualization purposes, all of these effects are accumulated: this allows you to move from the local level of a neighborhood in the grid to the global range of the feature.

The second step is quite important: instead of just evaluating the function on one point of the grid, you actually compute the slope of the function in the interval. Otherwise, you might end up confusing the effect of the feature of interest with that of other highly correlated features.

Figure 13-12 shows the ALE for the same simulated dataset used before, along with bootstrapped 90% confidence intervals. With uncorrelated features (first row), ALE does a great job of recovering the true effects. With correlated features (second row), the true effect of the second feature is recovered correctly, but some parts for the first feature still display some bias; nonetheless, ALE still does a better job than PDPs.

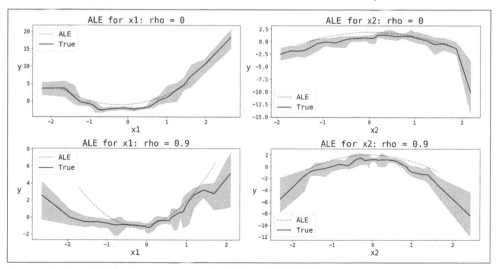

Figure 13-12. ALE for the same simulated data (90% CI)

Key Takeaways

These are the key takeaways from this chapter:

Holistic storytelling in ML
> In its most common usage, the act of storytelling in ML comes after you've developed your model and faced your stakeholders. The holistic approach presented in this chapter supports a vision where your scientist persona creates and iterates through stories that help you create a good predictive model, and then switches to the more traditional salesperson persona.

Ex ante storytelling
> Ex ante storytelling starts by creating stories or hypotheses about what drives the outcome you aim to predict. These are then translated to features through a multistep feature engineering stage.

Ex post storytelling
> Ex post storytelling helps you understand and interpret the predictions coming from your model. Techniques like heatmaps, partial dependence plots, and accumulated local effects should help you tell a story about the role that different features have on your outcome. Feature importance provides a way to rank them.

Structure the storytelling into steps
> At least at the beginning, it's good to put some structure on your storytelling toolkit, both from an ex ante and ex post point of view.

Further Reading

I discuss first- and second-order effects in *Analytical Skills for AI and Data Science*.

Rolf Dobelli's *The Art of Thinking Clearly* (Harper) is good if you want to gain some knowledge of the many biases and heuristics that are present in human behavior. These can greatly enrich the set of hypotheses for your specific problem.

On the problem of feature engineering, from a data transformation point of view, there are several comprehensive references out there. You can check out Alice Zheng and Amanda Casari's *Feature Engineering for Machine Learning* (O'Reilly), Sinan Ozdemir's *Feature Engineering Bookcamp* (Manning), Soledad Galli's *Python Feature Engineering Cookbook*, 2nd ed. (Packt Publishing), or Wing Poon's "Feature Engineering for Machine Learning" (*https://oreil.ly/Zg3EI*) series of blog posts.

I adapted Figure 13-5 from Figure 2.7 in *An Introduction to Statistical Learning with Applications in R*, 2nd ed. by Gareth James et al. (Springer) and available online from the authors (*https://oreil.ly/LZPDX*). This book is highly recommended if you're more interested in gaining some intuition than in understanding the more technical details.

On ML interpretability, I highly recommend Christoph Molnar's *Interpretable Machine Learning: A Guide for Making Black Box Models Explainable* (available online (*https://oreil.ly/FujJr*), independently published, 2023). Trevor Hastie et al., *The Elements of Statistical Learning: Data Mining, Inference, and Prediction*, 2nd ed. (Springer), has an excellent discussion on feature importance and interpretability for different algorithms (in particular, sections 10.13 and 15.13.2). Finally, Michael Munn and David Pitman give a very comprehensive and up-to-date overview of the different techniques in *Explainable AI for Practitioners: Designing and Implementing Explainable ML Solutions* (O'Reilly).

On ALEs, you can check the original article by Daniel W. Apley and Jingyu Zhu, "Visualizing the Effects of Predictor Variables in Black Box Supervised Learning Models" (August 2019, retrieved from arXiv (*https://oreil.ly/gbZlu*)). Molnar's account on ALE is very good, but this article can provide some further details into a somewhat less intuitive algorithm.

## CHAPTER 14

From Prediction to Decisions

According to a survey (*https://oreil.ly/Kl_7y*) done by McKinsey, 50% of their respondent organizations had adopted artificial intelligence (AI) or machine learning (ML) in 2022, a sharp 2.5x increase relative to 2017, but still lower than the peak reached in 2019 (58%). If AI is the new electricity (*https://oreil.ly/O_tsb*) and data the new oil (*https://oreil.ly/bU0xd*), why did adoption stall before the advent of large language models (LLMs) such as ChatGPT and Bard?[1]

While the root causes are varied, the most proximate cause is that the majority of organizations have yet to find a positive return on investment (*https://oreil.ly/Stpro*) (ROI). In "Expanding AI's Impact With Organizational Learning" (*https://oreil.ly/izJb7*), Sam Ransbotham and his collaborators argue that only "10% of companies obtain significant financial benefit from artificial intelligence technologies."

Where does this ROI come from? At its core, ML algorithms are predictive procedures, so it's natural to expect that most value is created by improved decision-making capabilities. This chapter goes into some of the ways that predictions improve decisions. Along the way, I will present some practical methods that will help you move from prediction to improved decision making.

Dissecting Decision Making

Prediction algorithms attempt to circumvent uncertainty, and doing so is extremely important in improving our decision-making capabilities. For instance, I can try to predict tomorrow's weather in my hometown for the pure pleasure of doing so. But

---

1 One may even ask if LLMs are really going to change the trend of adoption in a significant way. I believe that the fundamentals haven't really changed *yet*, at least until machines reach artificial general intelligence (AGI). But I'll discuss this topic in Chapter 17.

the prediction itself facilitates and improves our ability to make better decisions in the face of this uncertainty. It's not hard to find many use cases where different people and organizations would be willing to pay for this information (think farmers, party planners, the telecommunications industry, government agencies like NASA, etc.).

Figure 14-1 shows diagramatically the role that uncertainty plays in decision making. Starting from the right, once uncertainty is resolved, there is an outcome that affects some metric you care about. This outcome depends on the set of levers (actions) at your disposal and their interplay with the underlying uncertainty. For example, you don't know if it will rain today (uncertainty) and you care about being comfortable and dry (outcomes). You can decide to take your umbrella or not (levers). Naturally, if it rains, you're better off taking your umbrella (you're dry), but if it doesn't, the best decision is leaving it (you're more comfortable since you don't have to take it with you).

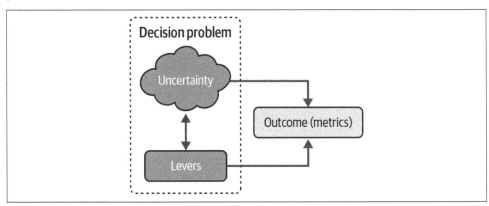

Figure 14-1. Decisions under uncertainty

In Table 14-1 I've assembled some common use cases in ML, where I highlight the roles that decision and uncertainty play, and some possible outcomes. Let's go through the first row, the case of health insurance claims processing (*https://oreil.ly/ 4V5Du*). Given a new claim, you must decide to review it manually or approve a payment, since a claim might be illegitimate. Illegitimate claims unnecessarily increase the insurer's costs, but review processes are often quite involved and take a substantial amount of time and effort. If you were able to predict correctly, you could lower prediction error and your costs, as well as increase customer satisfaction.

Table 14-1. Examples of ML use cases

Category	Use case	Decision	Uncertainty	Outcome
Service operations	Claims processing	Automatic payment versus review	Legitimate or not	Reduction of manual process (cost), higher customer satisfaction, lower fraud
Service operations	Staffing	Hire or relocate	Staff size depends on demand	Higher customer satisfaction, lower unused resources (cost)
Service operations	Proactive customer support	Call or not call a customer	Will a customer have a problem I can solve	Improve satisfaction and lower churn
Supply chain optimization	Demand forecasting	Manage inventory	Inventory depends on demand	Higher sales and lower depreciation costs
Fraud detection	Chargeback prevention	Approve or decline a transaction	Legitimate or not	Lower fraud-related costs, higher customer satisfaction
Marketing	Lead generation	Call or not call a potential customer	Will they buy or not	Higher sales efficiency
ML-based products	Recommender system	Recommend A or B	Will they buy or not	Higher engagement, lower churn

Thinking first about decisions and outcomes, and only then about ML applications, can take you a very long way in developing a strong data science practice at your organization.

Thinking about decisions and levers is a great way to find new ML use cases at the workplace. The process is:

1. Identify the *key* decisions made by your stakeholder (along with the relevant metrics and levers).

2. Understand the role of uncertainty.

3. Make a business case for building an ML solution.

Simple Decision Rules by Smart Thresholding

As opposed to regression, simple decision rules arise naturally in classification models in the form of *thresholding*. I will describe the case of a binomial model (two outcomes), but this same principle can be adjusted to the more general multinomial case. The typical scenario is something like this:

$$\text{Do}(\tau) = \begin{cases} A & \text{if } \hat{p}_i \geq \tau \\ B & \text{if } \hat{p}_i < \tau \end{cases}$$

Here $\hat{p}_i$ is the predicted probability score for unit i, and τ is a threshold chosen by you. The rule activates action A if the score is large enough, and action B otherwise. Note that a similar rationale applies if you replace the predicted probability with a predicted continuous outcome. However, the simplified structure inherent to classification settings allows you to include the cost of different prediction errors in your deliberation.

In a nutshell, everything boils down to a thorough understanding of false positives and negatives. In a binomial model, outcomes are usually labeled as positive (1) or negative (0). Once you have a predicted probability score and a threshold, units with a higher (lower) probability are predicted as positives (negatives). See the confusion matrix in Table 14-2.

Table 14-2. A typical confusion matrix

Actual/predicted	$\hat{N}(\tau)$	$\hat{P}(\tau)$
N	TN	FP
P	FN	TP

Rows and columns in the confusion matrix denote actual and predicted labels, respectively. As mentioned, predicted outcomes depend on the chosen threshold (τ). Thus, you can classify every instance in your sample as true negative (*TN*), true positive (*TP*), false negative (*FN*), or false positive (*FP*), depending on whether the predicted label matches the true label or not. Cells in the matrix denote the number of cases for each category.

Precision and Recall

Two common performance metrics in classification problems are the precision and recall:

$$\text{Precision} = \frac{TP}{TP + FP}$$

$$\text{Recall} = \frac{TP}{TP + FN}$$

Both metrics can be thought of as true positive rates, but each considers different universes.[2] Precision answers the question: *out of everything I said is positive, what percentage was actually positive?* On the other hand, recall answers the question: *out of everything that's actually positive, what percentage did I predict correctly?* When you

---

2 Note that in the ML literature, recall is commonly taken as the *true positive rate*.

use precision as your considerations, you are really thinking about the cost of a false positive; with recall, what matters is the cost of a false negative.

Figure 14-2 shows precision and recall curves for three alternative models trained on a simulated latent variable linear model for a balanced outcome. The first column shows a classifier that assigns a probability score by drawing random uniform numbers in the unit interval; this *random* classifier will serve as a baseline. The middle column plots precision and recall obtained from a logistic regression. The final column switches the predicted classes, on purpose, to create an inverse probability score where a higher score is associated with lower incidence rates.

You can readily see several patterns: precision always starts at the fraction of positive cases in your sample, and can be relatively straight (random classifier), increasing, or decreasing. Most of the time you get an increasing precision, since most models tend to outperform random classifiers and are at least somewhat informative of the outcome you want to predict. Though theoretically possible, a negatively sloped precision is highly unlikely.

Precision is better behaved, in the sense that it always starts at one and then decreases to zero, and only the curvature changes. A nice concave function (middle plot) is to be generally expected, and is also related to the fact that in healthy classification models, scores are informative of the probability of occurrence.

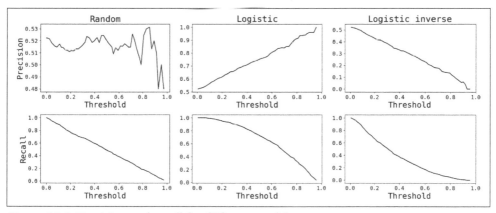

Figure 14-2. Precision and recall for different models

Example: Lead Generation

Take the example of a lead generation campaign, where you score leads to predict which will end in a sale. Your data consists of successful (sale) and failed (no sale) contacts for a historical sample of leads previously used by the telemarketing team.

Consider the simple decision rule to contact a customer if the predicted probability is higher than a threshold. An FN is a lead that would've become a sale had it been sent to the marketing team, and an FP is a lead that was incorrectly sent for contact, so it didn't end up in a sale. The cost of a false negative is the forgone revenue from the sale, and the costs of a false positive are any resources spent on the processing of a lead (for instance, if the hourly salary of a telemarketing executive is X, and each lead takes k minutes to be processed, the cost of each false positive is $kX/60$).

A simple *volume* threshold rule works like this: the sales team tells you how much volume (V) they can handle each period (day or week), and you send them the top V leads according to the estimated probability score. Clearly, by fixing the volume you also implicitly set the threshold of your decision rule.

Let's look at a simplified lead generation funnel (see also Chapter 2) to understand the effects of such a rule:[3]

$$
\text{Sales} = \underbrace{\frac{\text{Sales}}{\text{Called}}}_{(1)} \times \underbrace{\frac{\text{Called}}{\text{Leads}}}_{(2)} \times \underbrace{\text{Leads}}_{(3)}
$$
$$
= \underbrace{\text{Conv. Eff}(\tau)}_{\text{Precision}} \times \text{Call Rate(FTE)} \times \text{Leads}(\tau)
$$

Total sales depend on the conversion efficiency (1), call rate (2), and the volume of leads (3). Note that conversion efficiency and the volume of leads depend on the threshold you choose: in an idealized setting, conversion efficiency is equal to the *precision* of your model, and the number of leads depends on the scores distribution. On the other hand, the call rate depends on the number of full-time equivalent (FTE) or total employees of the sales team: large enough sales forces will be able to call every lead in the sample.

With this you can see why and when the volume rule may work. By sorting the leads on the probability score in a descending fashion, and contacting only the top V, you optimize conversion efficiency (since precision is an increasing function in predictive classification models). You also take care of idle resources in the telemarketing team: if you send more than they're able to handle, leads with lower scores won't be contacted in the current time window; if you send less, there will be idle sales agents.

Figure 14-3 plots the product of (1) and (3) as a function of the threshold set for the same simulated sample and the same three models used before.[4] Moving from right to left, you can see that lowering the threshold is always better from a total sales perspective, explaining why a volume rule usually works well for telemarketing teams.

---

3 I assume that the contact ratio is one, so every call ends in a contact. In applications this is usually not true, so not only does the funnel need to be expanded, but you may also need to adjust your model.

4 Sample size was normalized to 100 and the outcome is balanced, so there are only ~50 true positive cases.

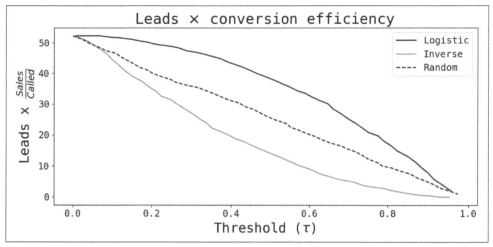

Figure 14-3. Optimizing total sales

One potential source of confusion arising from this figure is that it may suggest that you should set the threshold to zero (call every scored lead) instead of just following the volume rule. Put differently, should the sales team hire the exact number of FTEs that guarantee that the call rate is maximized and that all leads are contacted? The answer is negative: if the score is informative, leads with lower predicted scores are also less likely to convert, so the cost of an additional FTE (certain) will be larger than the (uncertain) benefit from the additional sale. The volume rule assumes that the size of the team is fixed, and then optimizes for the largest precision and sales, given this team size.

Confusion Matrix Optimization

The case of lead generation is somewhat atypical, because you effectively put zero weight to false negatives and focus only on optimizing precision. But this is not true for most problems (and even with lead generation, there's a case to be made to include false positives in the choice of a threshold). To see this, consider the case of fraud, where for any incoming transaction, you need to predict whether it's going to be fraudulent or not.

A typical decision rule blocks a transaction for large enough probability scores. False positives typically translate to infuriated customers (lower customer satisfaction and higher churn). On the other hand, a false negative creates a direct cost of fraud. This tension gives rise to interesting optimization problems for threshold selection.

The general idea is to find the threshold that *minimizes* the expected cost from incorrect predictions; alternatively, if you think you should also include the value from correct predictions, you can choose the threshold to *maximize* the expected profits. These can be expressed as:

$$E(\text{Cost})(\tau) = P_{FP}(\tau)c_{FP} + P_{FN}(\tau)c_{FN}$$

$$E(\text{Profit})(\tau) = P_{TP}(\tau)b_{TP} + P_{TN}(\tau)b_{TN} - \left(P_{FP}(\tau)c_{FP} + P_{FN}(\tau)c_{FN}\right)$$

where P_x, c_x, b_x denote the probability of a true or false positive or negative (x), and their associated cost or benefit, respectively. Probabilities are estimated using the frequencies in the confusion matrix as $P_x = n_x/\Sigma_y n_y$, and depend on the chosen threshold.[5]

Figure 14-4 shows sample estimates using the same simulated dataset as before; importantly, I assume a symmetric case where all costs and benefits have the same value (normalized to one). You can see that for cost (left) and profit optimization (right), the optimal threshold is ~0.5, as expected in a model with balanced outcomes and symmetrical cost/benefit structure.

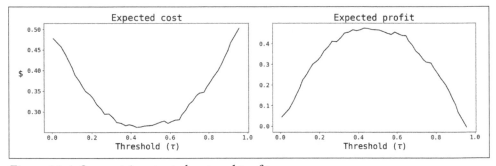

Figure 14-4. Symmetric expected cost and profits

Figure 14-5 shows the effect of doubling the cost of a false positive and negative on the optimal threshold. Directionally speaking, you would expect that increasing the cost of a false positive increases the threshold, as you put more weight on the precision of the model. Alternatively, a higher cost of a false negative lowers the optimal threshold since you put more weight on the recall.

---

5 While this is correct for the profit calculation, you may want to use the conditional probabilities, given a prediction error for the cost calculation. The chosen threshold doesn't change since this amounts to a rescaling of the objective function.

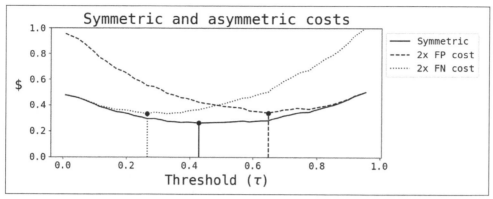

Figure 14-5. Asymmetric expected costs

You can use this method to find suitable thresholds that will transform your classification model into a decision rule. The process involves the following steps:

1. Train a classifier with good predictive performance.

2. For cost minimization, set suitable costs for prediction errors. Because of the structure of the problem, you need only have relative costs (such as, *the cost of a false negative is 3x that of a false positive*; that is, you can normalize everything with respect to one outcome).

3. A similar consideration applies for profit maximization.

4. These can be computed for different threshold values and optimized.

Key Takeaways

These are the key takeaways from this chapter:

Moving from prediction to decisions is critical if you want to find positive ROI for your data science practice.
> ML is a set of predictive algorithms that can, first and foremost, greatly improve your organization's decision-making capabilities.

Threshold decision rules abound in ML.
> Many regression and classification models give rise to simple decision rules that trigger actions if the predicted outcome is greater than, equal to, or lower than a predetermined threshold.

Decision rules in classification models.
> Because of the simplified outcome structure, classification models give rise to decision rules that can be easily optimized. One such optimization path takes into account the costs and benefits from different prediction outcomes (true and false positives or negatives). I showed how a simple volume-threshold rule arises

when you only care about the precision of your model, and the more complete case where false positives and negatives matter.

Further Reading

My book *Analytical Skills for AI and Data Science* goes in depth into many of the themes of this chapter. Importantly, I did not cover the practical problem of threshold optimization described here.

Ajay Agrawal et al., *Power and Prediction: The Disruptive Economics of Artificial Intelligence* (Harvard Business Review Press) strongly reinforce the point that the potential for AI and ML to disrupt the economy depends on their ability to improve our decision-making capabilities.

## CHAPTER 15

Incrementality: The Holy Grail of Data Science?

In the past (*https://oreil.ly/or6gY*) I've argued that *incrementality is the holy grail of data science*. This statement depends critically on the hypothesis that I've maintained throughout: that data science creates value by improving a company's decision-making capabilities. This chapter expands on this topic, but most importantly, I will present some techniques that should build some basic intuitions that will become handy if and when you decide to delve deeper. As usual, the topic is worthy of a book-length treatment, so I will provide several references at the end of this chapter.

Defining Incrementality

Incrementality is just another name for *causal inference* applied to decision-making analytics. If you recall from Figure 14-1, a typical decision comprises an action or lever, and an outcome that depends on the underlying uncertainty. If the lever *improves* the outcome, and you're able to isolate any other factors that might explain the change, you can say (with some degree of confidence) that it was incremental. For later reference, the action is also known as the *treatment*, following the more classical medical literature of controlled experiments, where some patients receive a treatment, and the remaining *control group* receives a placebo.

Causality is commonly defined by use of *counterfactuals*. As opposed to facts—something that we observe—counterfactuals attempt to provide an answer to the question: *what if I had followed a different course of action?* You can then say that an action has a causal effect on the outcome if the result is unique against all possible counterfactuals.

For instance, imagine you can pull a binary lever with only two possible actions, A and B (such as giving a price discount or not), and you observe an outcome Y (revenue). You end up giving a discount to all of your customers and find that revenue increases. Was the discount incremental on revenue? Alternatively, is this effect *causal*? To answer these questions, you need to estimate the counterfactual revenue, where every other factor is fixed and the only thing that changes is that you don't give a discount. The difference in these *potential outcomes* is the causal effect of the discount.[1]

By quantifying incrementality, you are able to identify and choose actions that put the company on an improving path. This is commonly associated with *prescriptive* analytics, as opposed to its descriptive and predictive counterparts. Most data scientists working on machine learning (ML) strictly focus on prediction and devote little to no time to think about causality, so you might wonder if this is really a critical skill to learn. Before moving on to more practical matters, I'll argue that it is.

Causal Reasoning to Improve Prediction

Even if you restrict your role as data scientist to prediction, you ought to care about causality in a very broad sense. As argued in Chapter 13, for engineering good, predictive features, you need to have some basic causal intuitions about the outcome you want to predict. This can be seen from the definition of supervised learning:

$$y = f(x_1, x_2, \cdots)$$

Given variation in your features and outcome $\{x_k, y\}$, the task is to learn the data generating process ($f()$). But this implicitly assumes a direction of causality from features to outcome. The process of feature engineering starts by formulating casual hypotheses of the type *a higher value of feature k increases outcome because....* Moreover, the predictive performance of your model may be negatively impacted if you include features that are spuriously correlated with the outcome, as explained in Chapter 10.

Causal Reasoning as a Differentiator

At the time of writing this book, GPT-4 and similar large language models (LLMs) are making us rethink the role of humans in many areas. Data scientists have heard about these risks before with the advent of automated machine learning (*https://oreil.ly/afagR*).

---

1 Just in passing, note that in this example there are alternative counterfactual stories that could explain the higher revenue. A very common one is peak seasonal sales, where customers are just more willing to spend more on your product.

But these technologies can make you more productive if you let the machines take care of everything that can be automated and devote your unique human capabilities on top of them. Even with the most recent advances, it seems safe to predict that for now, humans are uniquely suited to engage in causal reasoning by way of counterfactuals and building models of how the world works. Chapter 17 discusses this topic in detail.

Improved Decision Making

There is also the problem of how you create value for your organization. As I've argued throughout this book, data scientists are uniquely endowed with skills to improve a company's decision-making capabilities. If you follow this route, incrementality *is* the holy grail and there's no way you can escape thinking about causality.

But this route also requires you to rethink your role as data scientist, augmenting it from just prediction to enhanced decision making (where prediction plays an important, but secondary, role).

A typical scenario is the launch of a new feature or new product. When you launch a new feature, you'd better have an outcome or metric that you are trying to optimize. For instance, you may care about customer engagement, as measured by activity time or page visit frequency. If you're able to show that the feature was incremental on that metric, you can recommend expanding its use or augmenting it. Alternatively, if you don't find it to be incremental, or even worse, that the metric deteriorated, the best course of action is to roll the feature back.

The launch of new products adds the more interesting concept of *cannibalization*. For example, when Apple (*https://oreil.ly/QarTm*) decided to launch the iPhone, the sales of the iPod dropped significantly and were thus cannibalized. Similarly, the streaming business for Netflix (*https://oreil.ly/Zu5jM*) eventually displaced and cannibalized the original online DVD rental business. A somewhat different final example is the case of Starbucks (*https://oreil.ly/BCgCA*) opening a new store that may cannibalize the sales of neighboring stores. In all of these cases, estimating the incrementality for the new product or stores can have a deep impact on the company's P&L and decision-making capabilities.

Confounders and Colliders

Chapter 10 mentioned confounders and bad controls as examples of how things can go wrong with linear regression. Mastering these concepts is again of key practical importance when dealing with causality. I'll now review these concepts and highlight some places where you should direct your attention when thinking about incrementality.

One very useful tool to think about causality is directed acyclic graphs (DAGs). A *graph* is a set of nodes and links between the nodes. In this setting, nodes represent variables, and links denote causal relationships. When links are interpreted directionally, the graph becomes *directed*. For instance, if x causes y, there will be a directed link $x \rightarrow y$. The word *acyclic* precludes the existence of loops; if $x \rightarrow y$, it can't be that $x \leftarrow y$, so causal relationships are unidirectional. Judea Pearl, a computer scientist and Turing Award winner for his work on Bayesian networks, developed and popularized an approach to causal analysis using DAGs. Given your data and your DAG, the question is whether you can *identify* a specific causal effect. Identification is different from *estimation*, which uses statistical techniques to compute the sample estimate.[2]

Figure 15-1 shows DAGs for the simplest cases of a confounder and a collider where there's no causal effect from x to y. The left DAG shows that there are two causal relations ($c \rightarrow x$, $c \rightarrow y$), so c is a common cause for both x and y. On the right, there are also two causal relations ($c \leftarrow x$, $c \leftarrow y$), so c is a common effect.

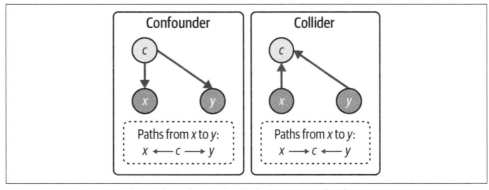

Figure 15-1. DAGs with confounder and collider: no causal effect

Confounder bias arises when two possibly unrelated variables (x, y) have a common cause (c). If you run a regression of y on x, *without controlling for c*, you will find that they are spuriously correlated. If the confounder is observed, all you need to do is condition on the confounder, and the causal relation, if any, will be identified. The problem arises with *unobserved* confounders, since by definition you can't control for them. In this case, you won't be able to identify the causal effect, if there is one.

A *collider* is a common effect of two variables, and is a typical example of a bad control, in the sense that including it in your regression will bias your estimates. If you

2 The DAG approach to identification is popular among computer scientists and epidemiologists, and the *potential outcomes approach* is most popular among statisticians and economists. I will talk more about the latter in what follows.

run a regression of y on x and control for c, you will find a spurious relation that doesn't exist.

To get a sense of what happens, I simulate the following data generating processes for a confounder (note that there's no causal effect from x to y):

$$c \sim N(0, 1)$$
$$\epsilon_x \sim N(0, 1)$$
$$\epsilon_y \sim N(0, 2)$$
$$x = 10 + 0.5c + \epsilon_x$$
$$y = -2 + 3c + \epsilon_y$$

Similarly, the data generating processes for a collider are (again, there's no causal effect from x to y):

$$\epsilon_x \sim N(0, 1)$$
$$\epsilon_y \sim N(0, 2)$$
$$\epsilon_c \sim N(0, 0.1)$$
$$x = 10 + \epsilon_x$$
$$y = -2 + \epsilon_y$$
$$c = 5 - 2x + 10y + \epsilon_c$$

I then run a Monte Carlo (MC) simulation where I estimate linear regressions of y on x with and without controlling for c. I report the estimated coefficient for the feature x and 95% confidence intervals in Figure 15-2.

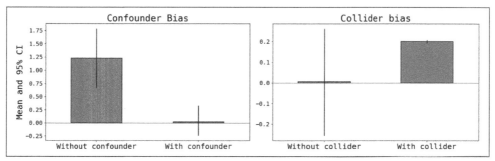

Figure 15-2. Confounder and collider bias (parameter estimate and 95% confidence interval)

For the confounder case, not controlling for c creates a statistically significant spurious correlation that would incorrectly indicate that x and y are related (even worse, you might end up concluding that x causes y). Notably, this correlation disappears once you include the confounder in the regression, leading to a correct inference about the nonexistent relation.

With the collider, the opposite happens: since it's a bad control, *excluding* it from the regression allows you to estimate a statistically insignificant effect of x on y. If you mistakenly think that c should be included as a feature, you end up concluding that there's a causal effect when there is none.

Both of these biases are pervasive in applications and unfortunately depend critically on your causal model for the outcome. Put differently, before attempting to estimate a causal effect, you must come up with a model (a DAG) for your outcome. Only then can you decide whether your available data is sufficient for identification for a given causal effect. Specifically, you must control for any confounders and ensure that you don't control for colliders (and sometimes these two considerations clash with each other, since a variable might be a confounder and a collider at the same time).

This process is often called the *back-door criterion*: with confounders you have to close any back doors by controlling for them, and with colliders the opposite applies; otherwise, you open those back doors and can't identify the causal effect.[3]

Another practical problem that arises has to do with proxy confounders. As already mentioned, unobserved confounders preclude identification of a causal effect, so you might be tempted to use proxy variables that are somewhat correlated with the confounder. The hope is that you can still estimate a causal effect using these less-than-optimal substitutes. Unfortunately, the answer is not good: the extent of the bias depends critically on the strength of the correlation. Figure 15-3 shows this for an MC simulation for the case of a confounder and a true causal effect of x on y.[4]

---

3 Note that the back-door criterion also includes a condition to not control for the descendants of the treatment (the variable that causes an outcome).

4 The DGP is essentially the same as before, but I introduced two changes: I draw c, proxy $\sim N(0, \Sigma(\rho))$ to allow for different correlation coefficients between the true unobserved confounder (c) and the observed proxy, and I model the outcome as $y = \hat{a}2 + 3c - 2x + \epsilon_y$ so that there is a causal effect from x to y.

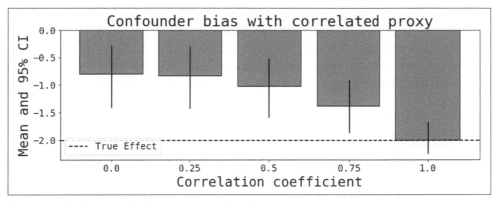

Figure 15-3. Confounder bias with correlated proxies

Selection Bias

Selection bias is a very important concept for causal analysis, but it has different meanings (*https://oreil.ly/TGxkr*) to different schools of thought. For statisticians and economists, it's associated with selection *into* the treatment, and for computer scientists it refers to a post-treatment selection that changes the sample of respondents; the former is a kind of confounding bias, and the latter produces a completely different DAG (better associated with survivorship bias, as discussed in Chapter 6). In this section I'll refer to the former (selection into the treatment), which is commonly associated with the potential outcomes literature. I will now introduce this notation.[5]

The idea of potential outcomes is closely related to counterfactuals. Consider the case of a binary treatment (D), where each unit i either gets it ($D_i = 1$) or not ($D_i = 0$). There's a unique potential outcome associated with each level of the treatment, denoted by Y_{1i} or Y_{0i}, corresponding to getting or not getting the treatment, respectively. For each unit we observe one and only one of these potential outcomes, denoted by Y_i; the other potential outcome is counterfactual so you don't observe it. The relation between the observed outcome and the potential outcomes can be summarized by:

$$Y_i = \begin{cases} Y_{1i} \text{ if } D_i = 1 \\ Y_{0i} \text{ if } D_i = 0 \end{cases}$$

Alternatively, $Y_i = Y_{0i} + (Y_{1i} - Y_{0i})D_i$, which maps quite neatly to the structure of a linear regression of the outcome on the treatment dummy variable and an intercept.

---

5 The distinction between the different types of selection bias is important. As I will show later, randomization precludes selection into the treatment, but doesn't solve the problem of post-treatment selection.

One advantage of thinking about causality in terms of potential outcomes is that the problem is essentially one of missing data. Table 15-1 shows one example, where each row denotes a customer. You only observe Y and D, from which you can immediately fill out the potential outcomes using the above logic. Were we able to observe each counterfactual outcome, we would be able to estimate the causal effect.

Table 15-1. Potential outcomes and missing values

	Y	Y0	Y1	D
1	6.28	6.28	NaN	0
2	8.05	8.05	NaN	0
18	8.70	NaN	8.70	1
7	8.90	NaN	8.90	1
0	9.23	9.23	NaN	0
16	9.44	NaN	9.44	1

To provide an example, suppose I want to estimate whether providing a GitHub repo with accompanying code for the book is incremental to the book's sales. My intuition is that knowing that there's available code increases the likelihood of a purchase, either because potential customers think the book is of higher quality, or because they know that the learning path is easier with code, I'd like to quantify the effect, since creating a code repo is costly. I will communicate and make it available to a sample of visitors to my web page ($D_i = 1$); to the remaining visitors I don't make it available ($D_i = 0$). The outcome is a binary variable, denoting a sale ($Y_i = 1$) or no sale ($Y_i = 0$).

For each unit i, $Y_{1i} - Y_{0i}$ is the causal effect of providing code. Since only one of these is observed for each unit, we need to estimate it using the sample of those who get and don't get the treatment. One natural way to estimate it is the *observed difference in means*: $E(Y_i|D_i = 1) - E(Y_i|D_i = 0)$. In practice, you replace the expectations with sample moments to get $\bar{Y}_{D_i = 1} - \bar{Y}_{D_i = 0}$, explaining why I say it's *observed*.

The bad news is that the observed difference does not estimate the causal effect in the presence of selection bias:

$$\underbrace{E(Y_i|D_i = 1) - E(Y_i|D_i = 0)}_{\text{Observed Difference in Means}} = \underbrace{E(Y_{1i} - Y_{0i}|D_i = 1)}_{\text{ATT (casual effect)}} + \underbrace{E(Y_{0i}|D_i = 1) - E(Y_{0i}|D_i = 0)}_{\text{Selection Bias}}$$

This decomposition is quite handy because it shows that in the presence of selection bias, the observed difference in means will deviate from the causal effect of interest, commonly denoted by the *average treatment effect on the treated* (ATT). The ATT answers the following question: looking only at those who received the treatment, what is the expected difference in outcomes between what they got and what they would've gotten had they not received the treatment? The second outcome is counterfactual, so the difference provides the causal effect on them.[6]

The third term represents selection bias and shows why the observed difference in means may deviate from the causal effect. To explain what this means, I will now use the following notation:

$$\text{Selection Bias} = \underbrace{E(Y_{0i}|D_i = 1)}_{A} - \underbrace{E(Y_{0i}|D_i = 0)}_{B}$$

Going back to the example, you can think of the code repo as a costly lever for a company (in this case, me) that can be assigned to everyone, or assigned selectively. Figure 15-4 shows the two types of selection bias. When there's positive (negative) selection, the causal effect tends to be overestimated (underestimated).

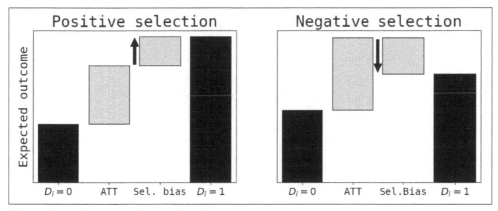

Figure 15-4. Positive and negative selection

Let's start with positive selection, which happens if I give the treatment to those that are already *more likely* to purchase the book. Alternatively, the probability of a sale is from the upstart higher for those who get the repo, independent of the incrementality of the lever. This means that $A \geq B$, overestimating the casual effect. A similar argument shows that with negative selection $A \leq B$, and the casual effect is underestimated.

Selection bias is pervasive in observational data. Either you (or someone from the company) selected the participants in the treatment, or the customers self-selected themselves. The code repo example is typical of selection by the company, but

---

6 Note that there are alternative casual effects you can estimate, namely the average treatment effect (ATE) or the average treatment effect on the untreated (ATU). I provide references at the end of this chapter.

self-selection is also very common. In Chapter 4 I introduced the idea of *adverse selection* where the riskiest customers—in terms of not being able to repay a loan—are also more willing to accept the offer. Adverse selection is a common example of self-selection.

 A thorough understanding of selection bias in your specific use case can take you very far in your quest to understand and estimate causal relations. Whenever you are looking at incrementality, ask yourself if there's any type of selection bias possible. This means that you have to think hard about the selection mechanism into the treatment you're analyzing.

Luckily, checking for selection bias is conceptually straightforward: take a set of *pretreatment* variables X and compute the difference for those in the treatment and control groups. Pretreatment variables are those that may affect selection into the treatment. Chapter 6 showed how lifts can be used, but for statistical reasons it's more common to use difference in means instead of a ratio (as this gives rise to a standard t-test).

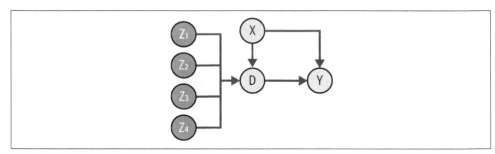

Figure 15-5. Selection bias and confounders

Figure 15-5 shows an example of a DAG that can be used to model selection bias. There are two sets of pretreatment variables $(Z_1, \cdots, Z_4, X)$, all affecting selection into treatment (D). Outcome (Y) depends on the treatment and X. Note that X is a confounder, and that variables Z_k create no bias if you control for the treatment. These other pretreatment variables may differ across treatment and control groups, but these differences don't create selection bias.

Unconfoundedness Assumption

With this in mind, it's time to introduce the main assumption needed for the identification of causal effects. This assumption goes by different names, such as *unconfoundedness, ignorability, conditional exchangeability, selection on observables,* and *conditional independence.*

The assumption means that the potential outcomes and selection into treatment are statistically independent, conditional on a set of observed controls:

$$Y_0, Y_1 \perp\!\!\!\perp D|X$$

This critical assumption has two alternative interpretations, one from the point of view of the decision-maker (owner of the selection mechanism) and the other from the data scientist's perspective.

Starting with the decision-maker, recall that selection into the treatment can either be done by the customer (self-selection) or by the owner of the treatment (you, or someone from your company). The assumption forbids that the decision-maker takes the potential outcomes into consideration.

For example, when discussing positive and negative selection, I was the owner of the selection mechanism, and it explicitly depended on whether I wanted to incentivize potential customers that were more or less likely to purchase organically. This is the same as saying that selection *depended on the potential outcomes*: if $Y_{0i} = 0$, the customer won't purchase the book without access to the repo, so I may wish to incentivize them (negative selection). A similar rationale applies for positive selection. Both of these cases might lead to a violation of unconfoundedness.

From the data scientist's perspective, you need to know in advance all of the relevant variables that could, in principle, affect the selection mechanism (you can then control for them and achieve conditional independence). I hope you can see why causal inference is so hard: not only do you need to have knowledge of the correct model for your outcome (the DAG), but you also need to observe all relevant variables. The latter explains why the assumption is also called *selection on observables*. Any unobserved confounders will result in selection bias. With observational data, both conditions are very hard to attain, which takes us to A/B testing.

Breaking Selection Bias: Randomization

Randomized controlled trials (RCT), or A/B testing as it's better known in the corporate jargon, are the quintessential method to estimate causal effects. The reason is that, by design, the unconfoundedness assumption is guaranteed: selection into the treatment depends only on the outcome of a pseudorandom draw, making it independent of the potential outcomes by construction.

Let's see how this works in practice. You need to first define the fraction of treated in a sample (p), which is usually set to one half in most common A/B test designs. You then draw from a uniform distribution (u) and define selection as:

$$D_i = \begin{cases} 1 & \text{if } u \geq p \\ 0 & \text{if } u < p \end{cases}$$

The following code snippet implements this random selection mechanism. The user provides the total sample size (n_total), the fraction of treated (frac_treated), and a seed for the random number generator, which allows for later replication. The outcome is a Boolean array that will indicate whether each unit in the sample is selected (True) or not (False).

```
def randomize_sample(n_total, frac_treated, seed):
    "Function to find a randomized sample"
    np.random.seed(seed)
    unif_draw = np.random.rand(n_total)
    bool_treat = unif_draw >= frac_treated

    return bool_treat
```

As mentioned, unconfoudedness is also called (conditional) *exchangeability*. In the example, had I randomized selection into the treatment, thanks to exchangeability I would expect the fraction of those who would organically buy the book to be the same in the treatment and control groups. Any incremental sales in one group must depend *solely* on the lever I provided. This is the beauty of A/B tests.

 When randomizing, you must ensure that the *stable unit treatment value assumption* (SUTVA) is satisfied. SUTVA has two requirements: (i) the treatment is the same for all individuals who get it (for instance, in drug trials, all patients must get the same equivalent dosage), and (ii) there is no interference between units, so the potential outcome for a unit doesn't vary with treatment assignment for other units.

The latter condition is often violated in online marketplaces, such as Uber, Lyft (*https://oreil.ly/Y3hWH*), or Airbnb. Suppose you want to test if a demand-side price discount improves revenue.[7] The treatment (discount) might reduce the available supply for the control group, creating an externality that affects their potential outcomes. In these cases, it's better to use randomization by blocks, where the sample is first split into mutually exclusive clusters, and treatment is randomized across clusters instead of across units.

---

7 Marketplaces have demand and supply sides. Examples of demand are the passenger in ridesharing, or guests for Airbnb.

Matching

As great as it is, A/B testing may not always be at your disposal, especially when the treatment is sufficiently costly. For instance, suppose you work for a telecommunications company that wants to know if installing an antenna (with all the required components) is incremental on its revenues. You can come up with an A/B test design where you randomize installing new antennas across geographical locations. Given a large enough sample size, this setup will allow you to estimate their incrementality. Needless to say, this test is just too costly to perform.

There are several techniques that will allow you to estimate causal effects with observational data, but they all depend on the critical unconfoundedness assumption to be satisfied. Here I just want to mention matching (and propensity score matching), because it nicely captures the intuition behind selection on observables, and how you can try to replicate randomization by finding a set of suitable units to be in the control group.

One way to think about randomization is that the treatment and control groups are *ex ante* equal, meaning that if you randomly choose one unit from each group and compare them with regard to any set of variables (X), they should be pretty much the same. A natural question is whether we can create a valid control group *ex post*, so that selection on observables applies. This is what *matching* attempts to do.

The matching algorithm works like this:

1. *Propose a DAG for your outcome and ensure that selection on observables is valid.* In practice, this means that you have a reasonable causal model for the selection mechanism and you can observe *all* pretreatment features X.

2. *Loop over all treated units:*

    a. *Find suitable individual control groups.* For each unit i, find the group of m units that are *closest* to i in terms of X. Denote it by $C(i)$. m is a metaparameter that controls the bias versus variance trade-off: most people use $m = 1$, potentially leading to low bias but large variance. You can increase the number and play with this trade-off.

    b. *Compute the average outcome for the control group.* Once you have a control group for unit i, you can compute the average outcome $\bar{y}_{0i} = (1/m)\Sigma_{j \in C(i)} y_j$.

    c. *Calculate the average treatment effect for unit i.* Compute the difference $\widehat{\delta}_i = y_i - \bar{y}_{0i}$.

3. *Calculate the average treatment effect on the treated.* The ATT is the average across all n_t individual treatment effects for units in the treatment group (N_T):

$$ATT = \frac{1}{n_t} \sum_{i \in N_T} \widehat{\delta}_{i_i}$$

I hope you like the simplicity and intuitiveness of the matching algorithm. The key insight is that each treated unit is matched to a control group that is most *similar* to it, with respect to any confounder. With continuous features, all you need to do is compute the Euclidean distance between *i* and all nontreated units *j*:

$$d_{ij} = \sqrt{\sum_k \left(x_{ik} - x_{jk}\right)^2}$$

What happens if you have mixed data, where your features can be continuous or categorical? In principle you can apply a general enough distance function.[8] But there's an alternative and very important result known as the *Propensity Score Theorem* (PST) that I will introduce now.

The *propensity score* is the probability that a given unit gets the treatment, conditional on some covariates or controls:

$$p(X_i) = \text{Prob}(D_i = 1 \mid X_i)$$

PST says that if unconfoundedness holds conditional on features X, it also holds if you condition on $p(X)$. The importance of this result is mainly computational: if you've already made the critical leap of *assuming* conditional independence using X, then you can use the propensity score to match treated with untreated units. The propensity score can be estimated with your favorite classification algorithm—such as a gradient boosting, a random, forest or a logistic classifier—that naturally takes care of mixed data.

---

8 For instance, see Kacper Kubara's post, "The Proper Way of Handling Mixed-Type Data. State-of-the-Art Distance Metrics" (*https://oreil.ly/gEn5R*).

Remember that unconfoundedness is an assumption that *cannot* be tested with any given dataset. You start with a DAG that captures your assumptions about any dependencies between the treatment, the outcome, and any other relevant controls. Everything that follows depends on this critical assumption.

Because of this criticality, it's always a good idea to discuss and document your identifying assumptions (your DAG) with your colleagues, data scientists or others. Many times, your business stakeholders can provide valuable insights into what drives the selection mechanism.

I'll sum up the propensity score matching algorithm now, skipping the common steps:

1. *Train a classification algorithm to estimate the probability of getting the treatment.* Using a sample of treated and untreated units, estimate $p(X_i)$.

2. *Match treated units using the propensity score.* For each treated unit i, compute the absolute differences in propensity scores for all untreated units:

$$d_{ij} = |\hat{p}(X_i) - \hat{p}(X_j)|$$

3. *Select the control group using the sorted differences.* Sort all differences in an increasing manner, and assign the top m to control group $C(i)$.

As intuitive as matching (and propensity score matching) is, it is computationally expensive since you have to loop through each treated unit, and for each of these you then have to loop over each untreated unit, and each feature, so you end up with a complexity of $O(n_t \times n_c \times k)$ in *Big O* notation. In the code repo (*https://oreil.ly/dshp-repo*) you'll find two versions of the algorithm, one using loops and one that uses Numpy's and Pandas' broadcasting capabilities, which considerably reduce the execution time.

To see the two in practice, I simulate a model similar to the one described previously, with two confounders that affect the selection probability and the outcome, and the true treatment effect is equal to two.[9] For the propensity score, I use two alternative algorithms: an out-of-the-box gradient boosting classifier (GBC) and logistic regression. I bootstrap 95% confidence intervals for each estimator. Figure 15-6 shows the results, where the horizontal axis of each plot shows what happens when you play with the size of the control group (m).

---

9 Details can be found in the code repo (*https://oreil.ly/dshp-repo*).

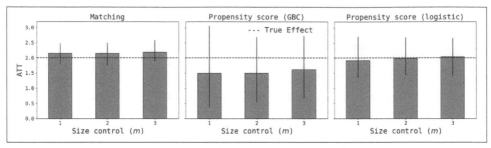

Figure 15-6. Results from matching and propensity score matching

It's clear that all methods correctly estimate the true causal effect, but propensity score matching with GBC slightly underestimates it (true estimate is still within 95% confidence intervals). Increasing the size of the individual control groups doesn't seem to have an effect, both in terms of bias and variance, for plain matching and propensity score matching with the logistic regression, but it slightly decreases the confidence intervals for GBC.

Machine Learning and Causal Inference

Although ML has enjoyed impressive growth in the past few years, it's safe to say that, other than A/B tests which are performed regularly at many organizations, causal inference is still quite niche. In this section I'll try to summarize some of the most recent developments that link these two fields of study.

Open Source Codebases

Just as the availability of open source libraries for ML removed some of the barriers to entry for practitioners, several new initiatives try to do the same for causal inference.

The causal inference research team at Microsoft has spun several projects that include EconML (*https://oreil.ly/8QMHp*), Azua (*https://oreil.ly/rowav*), and DoWhy (*https://oreil.ly/Ber5G*).

As the contributors to DoWhy explain (*https://oreil.ly/jaTr2*), their aim is to:

- Provide a modelling framework through casual graphs (DAGs)
- Combine the best of the DAG and potential outcomes approaches
- "Automatically [test] for the validity of assumptions if possible and [assess] the robustness of the estimate to violations"

The last objective is probably the most appealing to practitioners, since by providing the treatment, outcome, other data, and a causal model, you can get enough information about whether you have identification and a range of plausible estimates. As you might expect, automation is at the core of the research program led by Judea Pearl (*https://oreil.ly/JmYKa*) and the computer science crowd.

EconML is a Python library that aims at using state-of-the-art ML techniques to estimate causal effects. As the name suggests, the provided methods are "at the intersection of econometrics and [ML]." You can find some very recent methods that work under the unconfoundedness assumption, such as double machine learning, doubly robust learning, and forest based estimators. I will say more about this later.

Azua is a library that aims at using state-of-the-art ML methods to improve decision making. The problem is divided into two independent stages called *next best question* and *next best action*. The former is concerned about which data needs to be collected to make better decisions, and includes problems in missing value imputation and how informative different variables are for a given problem. The latter uses causal inference to provide optimal actions for well-defined objective functions.

CausalML (*https://oreil.ly/W2Vn8*) is another Python library, created by Uber. It includes several ML-based causal inference estimators for uplift modeling, such as trees and meta-learners. A similar library is pylift (*https://oreil.ly/Akxdj*).

To understand uplift modeling (*https://oreil.ly/3LMlX*), imagine that you train a cross-selling classifier that will predict which of your customers will purchase a given product in your company. Once trained, you can plot the distribution of scores, as in Figure 15-7, where I've divided all scored customers into three groups. Group A are customers with a high probability of a purchase. Customers in group B are less likely, and C are deemed highly unlikely to purchase.

Which customers should you target in your campaign? Many people decide to target group A, but these customers are most likely going to make an organic purchase, so you can use this costly incentive to target other customers. On the other side, group C are so unlikely that the incentive will be prohibitively costly. With this rationale, group B is a better candidate to be targeted.

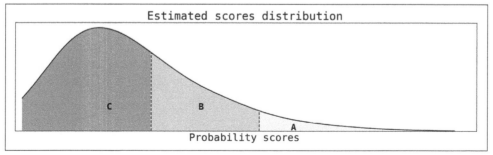

Figure 15-7. Distribution of cross-selling probability scores

The aim of uplift modeling is to formalize this intuitive discussion using the information on your treatment and control groups to estimate incrementality of the treatment.

Double Machine Learning

ML algorithms are great when the objective is to learn a general data generating process like $y = f(X)$. When using DAGs to describe a causal model, no mention is made of the functional form of the links, just their existence. Traditionally, causal effects are estimated using linear regression because of its simplicity and transparency. Double machine learning (DML) and similar techniques aim at using the increased predictive power and flexibility of nonlinear learners to estimate a causal effect.

To see how ML can improve the estimation of a causal effect, take the following partially linear model:

$$y = \theta D + g(X) + u$$
$$D = h(X) + v$$

As usual, the outcome depends on the treatment and some features, and the treatment also depends on the set of features (to create confounder or selection bias). The functions g and h are possibly nonlinear, the treatment effect is given by θ, and u, v are independent noise terms. Note that nonlinearity potentially kicks in only for the confounders, but these are not allowed to interact with the treatment.

The idea of the DML estimator is to use the power of nonlinear learners (such as random forests or gradient boosting) to learn each of these functions and estimate the treatment effect. Without going into the details, the process involves two critical concepts:

Orthogonalization
    As described in Chapter 10, orthogonalization consists of partialling out the effect of the covariates X on the outcome and the treatment. You use the desired flexible learner and regress the residuals to obtain the causal effect.

Sample splitting
    The sample is randomly split in halves, one used for training and the other for estimation and evaluation. This is necessary to avoid bias from overfitting, and provides some desirable large sample properties.

The algorithm works like this:

1. Randomly split the sample in two halves: $S_k, k = 1, 2$.

2. Using sample l, train your learners on S_l for both $g()$ and $h()$.

3. Using units i in sample $m \neq l$, estimate the residuals:

$$\hat{u}_i = y_i - \hat{g}(X_i)$$
$$\hat{v}_i = D_i - \hat{h}(X_i)$$

4. Calculate the estimator:[10]

$$\hat{\theta}(S_l, S_m) = \left(\frac{1}{n_m} \sum_{i \in S_m} \hat{v}_i D_i \right)^{-1} \left(\frac{1}{n_m} \sum_{i \in S_m} \hat{v}_i \hat{u}_i \right)$$

5. Average the estimates from each subsample:

$$\hat{\theta} = 0.5 \times \left(\hat{\theta}(S_l, S_m) + \hat{\theta}(S_m, S_l) \right)$$

In the code repo (*https://oreil.ly/dshp-repo*) you can find an implementation and results for simulations using linear and nonlinear data generating processes. Here I just wanted to show one avenue where ML has impacted causal inference by providing more powerful and general predictive algorithms.

Key Takeaways

These are the key takeaways from this chapter:

What is incrementality?
Incrementality is causal inference applied to estimating whether a change in a lever improved a business outcome.

Why care about incrementality (v.0)?
Under the assumption that data science creates value by improving our decision-making capabilities, incrementality is critical to understand which decisions are worthy of expanding and which should be rolled back.

---

10 Note that this expression isn't exactly what you would obtain from the Frisch-Waugh-Lovell procedure of regressing partialled out residuals. This expression is actually closer to an *instrumental variables* estimator (see the references at the end of this chapter). The creators of double machine learning present another estimator that follows more closely the FWL logic (see their Section 4).

Why care about incrementality (v.1)?

Even if improved decision making is not a top priority for you or your team, having a broad understanding of causality should help you improve the predictive performance of your ML models.

Approaches to causality

Generally speaking, there are two alternative (and complementary) approaches to identification and estimation of causal effects: the DAG and potential outcomes methodologies. The former takes advantage of graphs (and the do-calculus) to find conditions for identification. The latter transforms the problem into one of missing data and selection mechanisms, since at any given time, only one potential outcome can be observed for each unit.

Confounders and colliders

Confounders are common causes, and colliders are common effects for a treatment and an outcome. Not conditioning for a confounder *opens a back door* and results in biased causal estimates. Alternatively, a collider is an example of a bad control in the sense that including it as a feature in your model (or more generally, conditioning on it) will also open a back door and create bias.

Selection bias

For statisticians and economists, selection bias is a type of confounder bias applied to selection *into* the treatment. For epidemiologists and computer scientists, it refers to selection into a sample *after* the treatment was administered. Randomization, in the form of RCTs or A/B tests, solves the former but not the latter.

Randomization and matching

By randomizing selection into the treatment, you effectively break selection (into the treatment) bias. This explains why A/B tests have become an industry standard whenever the option is available. With observational data there are many techniques that can be used to estimate causal effects, but they all rely on the unconfoundedness assumption to be valid. Here I discussed matching and propensity score matching only.

Further Reading

In my book, *Analytical Skills for AI and Data Science*, I discuss in depth the relevance of incrementality and causality for prescriptive data science. A similar view can be found in Ajay Agrawal et al., *Prediction Machines: The Simple Economics of Artificial Intelligence* (Harvard Business Review Press) and the more recent *Power and Prediction: The Disruptive Economics of Artificial Intelligence* (Harvard Business Review Press).

An introductory treatment of casual inference can be found in Chapter 9 (*https://oreil.ly/j2JfH*) of *Data Analysis Using Regression and Multilevel/Hierarchical Models* by Andrew Gelman and Jennifer Hill (Cambridge University Press).

If you're interested in the DAG approach to causality, an introduction can be found in Judea Pearl and Dana Mackenzie, *The Book of Why: The New Science of Cause and Effect* (Basic Books). A more technical treatment can be found in Pearl's *Causality: Models, Reasoning and Inference*, 2nd ed. (Cambridge University Press). The former is better suited if you first want to gain some intuition, and the latter provides an in-depth presentation of DAGS and the do-calculus. Of critical importance are the back- and front-door criteria for identification.

The potential outcomes approach has been championed by economists and statisticians. *Mostly Harmless Econometrics: An Empiricist's Companion* by Joshua Angrist and Jorn-Steffen Pischke (Princeton University Press) is a great reference if you're interested in understanding selection bias and the many facets of linear regression as compared to other methods, such as the matching estimators discussed in the chapter. You can also find a complete treatment of instrumental variables, discussed in a footnote of the DML estimator.

Causal Inference for Statistics, Social, and Biomedical Sciences: An Introduction by Guido Imbens and Donald Rubin (Cambridge University Press) provides a thorough introduction to the subject from a potential outcomes perspective, also known as Rubin's causal model (Donald Rubin originally formalized and developed the theory). This is a great reference if you want to understand the role that selection mechanisms play. SUTVA is also discussed in great detail.

In recent years, several authors have tried to make the best of both approaches. On the economist's side, Scott Cunningham's *Causal Inference: The Mixtape* (*https://oreil.ly/mlTOy*) (Yale University Press) and Nick Huntington-Klein's *The Effect: An Introduction to Research Design and Causality* (*https://oreil.ly/DewAm*) (Chapman and Hall/CRC) discuss several methods for identification and estimation, and also provide clear introductions to DAGs.

While Miguel Hernan and James Robins are very respected in the DAG literature, their book *Causal Inference: What If* (CRC Press) uses potential outcomes to introduce causality and counterfactuals, and derive many important results using DAGs.

Guido Imbens, who shared the Nobel Prize in economics (*https://oreil.ly/8p3Yr*) with David Card and Joshua Angrist in 2021, has been involved in several discussions with Judea Pearl on the relative usefulness of both approaches. You can find his view and review in "Potential Outcome and Directed Acyclic Graph Approaches to Causality: Relevance for Empirical Practice in Economics" (working paper (*https://oreil.ly/OcAm8*), 2020). You might also be interested in reading Judea Pearl's response (*https://oreil.ly/tz8Hl*).

Also, if you're interested in how these different schools of thought evolved and their views, you can check the ungated special issue (*https://oreil.ly/MXYlp*) of *Observation Studies* 8, no. 2 (2022). It has interviews with Judea Pearl, James Heckman (another Nobel Prize winner in economics), and James Robins (an epidemiologist who has led the research on causal inference through structural modeling) on their views about the subject and the different approaches.

Carlos Cinelli et al., "A Crash Course in Good and Bad Controls" (*Sociological Methods and Research*, 2022, available online (*https://oreil.ly/TqTkX*)), is a systematic discussion on the problem of bad controls.

Elias Bareinboim et al., "Recovering from Selection Bias in Causal and Statistical Inference" (*Proceedings of the AAAI Conference on Artificial Intelligence* 28, no. 1, 2014, also available online (*https://oreil.ly/ZCxGS*)) discuss selection bias from the point of view of post-treatment sample selection. On this topic, you can also read Miguel Hernan's discussion of different types of bias (*https://oreil.ly/B6rey*), and Louisa H. Smith's paper "Selection Mechanisms and Their Consequences: Understanding and Addressing Selection Bias" (*Current Epidemiology Reports* 7, 2020, also available online (*https://oreil.ly/uqNR4*)).

Trustworthy Online Controlled Experiments by Ron Kohavi et al. (Cambridge University Press) discusses many important topics in A/B test design, including the problem of interference or SUTVA violations. You can also check Peter Aronow et al., "Spillover Effects in Experimental Data," in J. Druckman and D. Green, eds., *Advances in Experimental Political Science* (Cambridge University Press, arXiv (*https://oreil.ly/ZrQQa*)).

Matheus Facure's *Causal Inference in Python* (O'Reilly) provides an overview of many of the topics discussed here in a book-length treatment. You can also check online his "Causal Inference for the Brave and True" (*https://oreil.ly/IgsQE*).

On uplift modeling, you can check Shelby Temple's "Uplift Modeling: A Quick Introduction" (*https://oreil.ly/uqdHd*) post (*Towards Data Science*, June 2020). Chapter 7 in Eric Siegel's *Predictive Analytics: The Power to Predict Who Will Click, Buy, Lie, or Die* (Wiley) has an introduction to the topic for the general public.

Jean Kaddour et al., "Causal Machine Learning: A Survey and Open Problems" (2022, arXiv (*https://oreil.ly/OBIUu*)), provides an up-to-date summary of many important topics aside from ML and causality that were not discussed in this chapter.

If you want to learn about double machine learning, the original article (*https://oreil.ly/TIcnB*) is written by Victor Chernozhukov and his coauthors, "Double/Debiased Machine Learning for Treatment and Structural Parameters" (*Econometrics Journal* 21, no. 1, 2018). I also found useful Chris Felton's lecture notes (*https://oreil.ly/3ZkfG*) and Arthur Turrell's "Econometrics in Python Part I— Double Machine Learning" post (*https://oreil.ly/89gBR*). There are Python and R packages (*https://oreil.ly/3M6bU*). The EconML (*https://oreil.ly/Ks5RT*) package also has methods to estimate DML.

## CHAPTER 16

A/B Tests

Chapter 15 described the importance of randomization to estimate causal effects, when this option is actually available to the data scientist. A/B tests use this power to improve an organization's decision-making capabilities in a process analogous to *local optimization*.

This chapter describes A/B tests and should help you navigate the many intricacies of a relatively simple procedure for improved decision making.

What Is an A/B Test?

In its simplest form, an *A/B test* is a method to evaluate which one of two alternatives is better in terms of a given metric. *A* denotes the default or baseline alternative, and *B* is the contender. More complex tests can present several alternatives at the same time to find the best one. Using the language from Chapter 15, units that get *A* or *B* are also called *control* and *treatment* groups, respectively.

From this description you can see that there are several ingredients in every A/B test:

Metric
> Being at the heart of improved decision making, the design of A/B tests should always start by choosing the right metric. The techniques described in Chapter 2 should help you find a suitable metric for the test you want to implement. I'll denote this outcome metric with Y.

Levers or alternatives
> Once you define a metric, you can go back and think of the levers that most directly affect it. A common mistake is to start with an alternative (say, the background color of a button in your web page or app) and try to reverse engineer

some metric. I've seen this many times in practice, and it almost always leads to wasted time, team frustration, and inconclusive results.

Randomized selection

You must always define who gets access to each alternative. A/B tests are also called *randomized controlled trials* because, by design, selection into the treatment is random, thereby breaking any confounder or selection bias that may arise.

Decision Criterion

Each unit i participating in the experiment has an associated outcome, denoted by Y_i. At the end of the experiment, you have collected data on this metric for units in both groups, and your task is to decide if the new alternative beats the default or not.

There are several ways to pose this problem, but the most common is to compare the sample averages for both groups. The key difficulty here is that you need to disentangle signal from noise.

Figure 16-1 shows two typical scenarios. Each plot displays the outcome measurements for each unit across treatment and control groups (vertical lines), as well as the sample means (triangles). On the left, you have a *pure noise* scenario, where the distributions of outcomes are the same for treatment and control, but if you just compared the means, you would conclude that lever B was superior. On the right, the treatment shifted the distribution to the right, creating a real discrepancy between average outcomes.

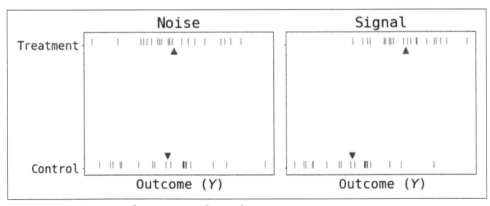

Figure 16-1. Disentangling noise and signal

Statistical tests allow you to formalize these intuitions. Typically, a null hypothesis is contrasted against an alternative, and you compute a test statistic with a known distribution. Denote by $\bar{Y}_k, k \in \{A, B\}$ the sample average for units in group G_k:

$$\bar{Y}_k = \frac{1}{N_k} \sum_{i \in G_k} Y_i$$

The most common criterion used in A/B tests is the following:

Keep lever k if $\bar{Y}_k - \bar{Y}_j > 0$ and the difference is statistically significant

Under this criterion, all you need to do is run a standard t-test that contrasts the null that there's no effect against an alternative hypothesis. Denote the difference in average outcomes by $\hat{\theta} = \bar{Y}_B - \bar{Y}_A$. A two-sided statistical test is:

$$H_0 : \hat{\theta} = 0$$
$$H_1 : \hat{\theta} \neq 0$$

H_0 denotes the null hypothesis that there's no difference in outcomes. Your goal is to *reject* this hypothesis with some degree of confidence; if you can't, you keep the default lever A (or not, since they are indistinguishable from the point of view of this specific metric).

Figure 16-2 shows how this is done in practice. The figure shows a theoretical distribution for your test statistic *under the null hypothesis of no effect* (note that it's centered at 0), which is usually taken to be a Student's t distribution. You compute your t statistic, and if it falls on the shaded area (rejection zone), you can reject the null at α significance level, which is usually set at 5% or 1%. This is the area of the shaded region in the plot, and is chosen to be small enough.

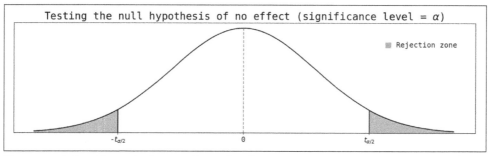

Figure 16-2. Deciding whether you keep the alternative treatment

I want to stop here to interpret what I just did. By choosing a small enough significance level, you essentially say: *if the null hypothesis is true, seeing such a large value for the test statistic is so unlikely that maybe my null was wrong.* Put differently, highly unlikely occurrences under the null are taken as evidence for rejecting the null. For instance, if you choose a 1% significance level, you should observe a test statistic that falls on the rejection region 1 in 100 times. But you got it in your experiment! Either you were extremely unlucky, or your null was wrong. You take the latter route, dismiss luck, and reject the null.

Let's run through an example using only 10 observations from the dataset on the left panel in Figure 16-1 (see Table 16-1).

Table 16-1. Outcome for first 10 units

IDs	Control	Treatment
0	0.62	0.82
1	1.07	0.23
2	0.56	2.47
3	−0.61	0.54
4	2.63	1.12
5	0.17	−0.40
6	0.94	−1.12
7	1.44	2.60
8	2.25	1.39
9	1.42	0.76
Mean	1.05	0.84

For these 10 units, the difference in mean outcomes is $\hat{\theta} = 0.84 - 1.05 = -0.21$. To compute the *t*-statistic, we first need the variance of the difference:

$$s_k^2 = \sum_{i \in G_k} (Y_i - \bar{Y}_k)^2 / (N_k - 1)$$

$$Var(\hat{\theta}) = Var(\bar{Y}_B) + Var(\bar{Y}_A) = \frac{s_B^2}{N_B} + \frac{s_A^2}{N_A} = 0.224$$

$$\text{t-stat} = \frac{\hat{\theta}}{\sqrt{Var(\hat{\theta})}} = -0.44$$

Is this *t*-statistic large enough to reject the null of no effect? We can use a table with critical values, or alternatively, directly compute the *p* value (the number of degrees of freedom is $N_B + N_A - 2$):[1]

---

1 *F* denotes the cumulative distribution function for the *t* distribution.

$$p \text{ value} = 2\big(1 - F(\,|\,\text{t-stat}\,|\,)\big) = 0.67$$

Under the null, there's a 67% probability of seeing a value at least as extreme as plus/minus 0.44. Since this is not *small enough* (usually < 5%), you can't reject the null hypothesis that this is pure noise. From a decision point of view, you stay with the default alternative.

You can also use linear regression to arrive at *the exact same* result. To do so, run the regression:

$$Y = \alpha + \theta D + \epsilon$$

Once $\hat{\theta}^{ols}$ is computed, you can use the p value that many packages precompute. Note that scikit-learn (*https://oreil.ly/nOe0Y*) *doesn't* compute p values, but you can use statsmodels (*https://oreil.ly/hRZKC*) to do so. In the code repo (*https://oreil.ly/dshp-repo*), I show you how to implement this manually, using statsmodel and SciPy's *t*-test method (*https://oreil.ly/apotw*).

Other than simplicity, linear regression also allows you to include other control variables (features), that may provide smaller confidence intervals. I will provide references at the end of this chapter.

Minimum Detectable Effects

I hope I've convinced you that this decision criterion is quite easy to implement using the following three-step process:

1. Fix a significance level (say 5%).
2. Compute the test statistic and the p value.
3. Reject the null of no effect if the p value is lower than the significance level.

I discussed similar threshold-based decisions in Chapter 14 where false positives and negatives arose naturally. It turns out that false positives and negatives play an important role in the design of A/B tests too.

In this context, a false positive arises if you incorrectly conclude there's an effect of the experiment; a false negative arises if you incorrectly conclude there's no effect.

As discussed, the significance level controls the probability of a false positive. When you reject the null (a *positive* because you say there is an effect), the probability that you made a mistake is given by the significance level (α). On the other hand, *statistical power* allows you to control the probability of a false negative, and is critical for the design of experiments.

Figure 16-3 shows two distributions: on the left, the distribution is centered at 0, assuming that there's no effect ($\theta = 0$). On the right, I plot another distribution under the assumption that *there is* a positive effect ($\theta^* > 0$). This second distribution will be used to discuss false negatives.

For a given significance level, the shaded area *FP* denotes the probability of a false positive, where you incorrectly reject the null, thereby concluding there is an effect when actually there's none. Now suppose that you conclude that there's *no* effect. This happens whenever your *t* statistic falls to the left of the critical value t_α.[2] For this to be a *false negative*, it must be that the true distribution is something like the one on the right, and the shaded area *FN* denotes the probability of a false negative for that distribution.

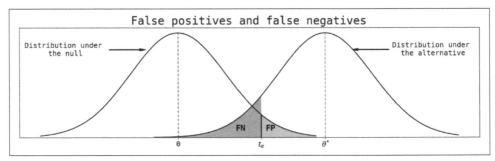

Figure 16-3. Understanding false positives and negatives

The *minimum detectable effect* (MDE) is the minimum effect of the experiment that you can detect for a given significance level and statistical power. It is given by the following formula, where $N = N_A + N_B$ is the total sample size of the experiment, $P = N_B/N$ is the fraction of treated units, and, as before, t_k are critical values from a t distribution:[3]

$$MDE = \left(t_\alpha + t_{1-\beta}\right)\sqrt{\frac{Var(Y)}{NP(1-P)}}$$

Why are MDEs so important for you? *Even if a true effect exists*, when it's smaller than the MDE, you'll be able to estimate it, *but* it will appear to be statistically insignificant. In practice this means that you run a test and conclude that the treatment had no incremental effect. The question is whether this is a true or a false negative. In *underpowered tests*, you won't be able to say if it's one or the other.

---

2 The subscript is now α instead of $\alpha/2$ because I'm considering a one-sided test now.

3 You can find a derivation here (*https://oreil.ly/C-rt9*).

As this discussion suggests, your objective when designing an A/B test is to achieve the *lowest* MDE possible. Small MDEs guarantee that, by design, you will be able to find equally small signals (the true effect) among all the noisy data you have.

Figure 16-4 shows the relationship between MDEs, sample size, and the variance of the outcome. For a fixed variance, increasing the sample size in your experiment lowers the MDE. Alternatively: *the larger the experiment, the better for estimating small effects.*

Now, fix a sample size and draw a vertical line across the different curves in the figure. The noisier your data (higher variance), the higher the MDE. The lesson is that with noisy data, you need even larger sample sizes to get comparable MDEs.

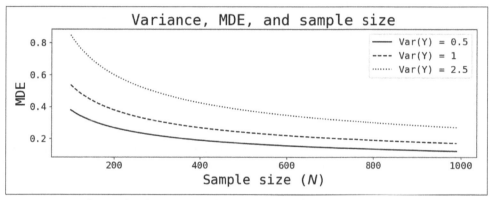

Figure 16-4. Relationship between MDE, variance, and sample size

 To summarize the key points from this section:

- You want to design tests with small MDEs.
- To do so you need to have larger experiments in terms of sample size.

As simple as this sounds, keep in mind that designing larger experiments may impact the operation at your organization. Many times you need to freeze any communications to participants in a test for several months, so large experiments also have a downside. I'll discuss this later when I talk about governance of experiments.

Example 16-1 shows how you can compute the MDE in Python. To find the critical values for the t distribution, the user must provide the statistical size (significance) and power (or use the default values). The number of degrees of freedom are usually a function of your sample size; here I'm setting it to $n - 1$, but for large enough sample sizes, the correction is unnecessary.

To find the critical values, you use the inverse function of the cumulative distribution function (CDF). In SciPy (*https://oreil.ly/Wusn7*) you can use the method `scipy.stats.t.ppf()`. Since I want the critical value on the right tail of the distribution for t_α, I need to subtract the significance level from one. A similar argument works for the second critical value ($t_{1-\beta}$), but now focusing on the left tail of the distribution.

Example 16-1. A Python script to compute the MDE

```
def compute_mde(sample_size, var_outcome, size=0.05, power = 0.85):
    # degrees of freedom: usually a function of sample size
    dof = sample_size - 1
    t_alpha = stats.t.ppf(1-size, dof)
    t_ombeta = stats.t.ppf(power, dof)
    p = 0.5
    den = sample_size*p*(1-p)
    MDE = (t_alpha + t_ombeta)*np.sqrt(var_outcome/den)
    return MDE
```

Many times you don't need the MDE, but rather the minimum sample size that is consistent with a desired MDE, to help you choose the right size of your experiment. Fortunately you can invert the function and solve for the sample size as a function of everything else; note that now you have to provide an MDE. Example 16-2 shows how to do it.

Example 16-2. A Python script to compute the minimum sample size

```
def compute_sample_size(mde, var_outcome, data_size, size=0.05, power = 0.85):
    # data_size is the number of subjects used to compute the variance of the outcome
    # (var_outcome)
    dof = data_size - 1
    t_alpha = stats.t.ppf(1-size, dof)
    t_ombeta = stats.t.ppf(power, dof)
    sum_t = t_alpha + t_ombeta
    p = 0.5
    sample_size = var_outcome/(p*(1-p))*(sum_t**2/mde**2)
    return sample_size
```

I will now discuss the choice of the remaining parameters.

Choosing the Statistical Power, Level, and P

It's common practice that you choose $\alpha = 0.05$ and $\beta = 0.15$. While you would like to have both be as small as possible, for a fixed MDE you need to trade off one for the other, which in practice means trading off the probabilities of a false positive and a false negative (see Figure 16-5). When designing your experiment, you can bring that into consideration and see what matters most for you. Just remember to interpret these values correctly: 5% is the probability of a false negative under the null, and 15% is the probability of a false positive under the alternative.

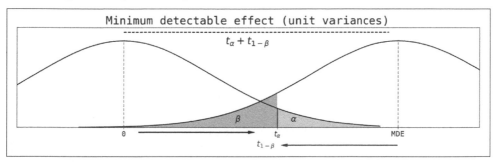

Figure 16-5. MDE, significance level, and power

To set the fraction of treated units (*P*), note that, all else equal, the MDE is minimized when *P* = 0.5, so this makes it a reasonable choice. In practice, this means that the treatment and control groups are equally sized.

Estimating the Variance of the Outcome

The last parameter you need is the variance of your outcome ($Var(Y)$). Many times, you can actually estimate this from your existing data. For instance, if your outcome is the average revenue per user, you can get a random sample of customers in your database and estimate the variance across those units.

There's another trick when the outcome is a binary variable, like a conversion rate. For instance, if your experiment is designed to see if a new feature improves a conversion rate, each individual outcome is $Y_i \in \{0, 1\}$, depending on whether it ends in a sale or not. You can model this as a Bernoulli trial with probability *q*, for which you know that the variance is $Var(Y) = q(1 - q)$. You can use the average of conversion rates from previous campaigns and substitute it for *q* in the equation to get an estimate.

Finally, you can always first run an A/A test. As the name suggests, units in both groups are presented with the default alternative *A*. You can then use the results from this experiment to estimate the variance of the outcome.

Simulations

Let's run some simulations to insure that all of these concepts are clear. I'll use the following simple data generating process for both simulations:

$$\epsilon \sim N\left(0, \sigma^2\right)$$
$$D \sim \text{Bernoulli}(p = 0.5)$$
$$y = 10 + \theta D + \epsilon$$

My first simulation uses $\theta = 0.5$, $\sigma^2 = 3$, so there's a small true effect (relative to the noisy data). The second simulation keeps the residual variance the same, but now there's no true effect ($\theta = 0$, $\sigma^2 = 3$). The sample size for each simulation is 500.

For the first simulation, I calculated the minimum sample size that allows me to detect the true effect ($N(MDE = 0.5) = N^* = 346$). I then created a grid of sample sizes that go from 50% to 150% of this size, and for each sample size, I draw with replacement, from the overall sample, 300 subsamples of this size. For each of these I estimate a linear regression that includes the intercept and the dummy variable, and flag a positive (negative) whenever the p value for the dummy variable is lower (higher) than the 5% significance level, as I would if the experiment was real. Finally I compute the true positive and false negative rates by averaging out the flags.

Figure 16-6 plots the true positive (TPR) and false negative (FNR) rates for the first simulation, along with the power used to calculate the minimum sample size. As you would expect, the TPR is increasing and the FNR is decreasing in the sample size: larger experiments have lower prediction errors.

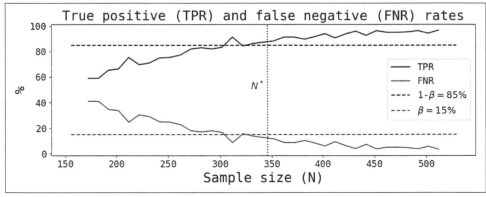

Figure 16-6. True positive and false negative rates: $\theta = 0.5$

The most important finding is that both lines cross the respective thresholds $\beta = 15\%$ when the sample size is as large as the minimum size that I got using the MDE formula. To repeat myself, this means that *even if your experiment had an effect, unless you have a large enough sample size, you will dismiss it as statistically insignificant.* What is *large enough* in the simulation? The sample size that lets me detect the true effect. This shows the beauty of the MDE formula, and hopefully it also helps you grasp the intuition behind it.

Figure 16-7 shows the results for the second simulation where there's no effect $\theta = 0$. Using the same decision criterion, if the *p* value is smaller (larger) than 5%, I flag a result as a false positive (true negative). The figure should help reinforce your understanding of significance levels and *p* values. In this case, around 5% of the time you incorrectly conclude that the experiment had an effect.

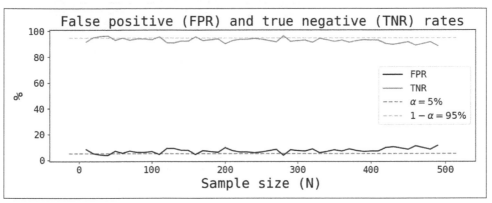

Figure 16-7. False positive and true negative rates: $\theta = 0$

Example: Conversion Rates

Let's run through a more realistic example to see if every concept is clear. You want to design an A/B test to see whether a different wording for automated email communications can improve upon the baseline 4% conversion rate that the company currently has.

Figure 16-8 shows the conversion rate (baseline + MDE) that you would be able to detect if you used a one thousand, one million, or one billion sample size. With 1K customers in the test, you can only detect incremental changes of at least 3.3 pp. For instance, you won't be able to detect a highly successful test where the new message creates a 5.5% conversion rate! If you only have access to a 1K size sample, the recommendation would be to *not* run the test, since you will only be able to detect unrealistically high incremental effects.

If instead you have access to 1M customers, the MDE is now 0.001, so any conversion rate larger than 4.1% would be detected. This sounds quite promising now, but the sample size could be prohibitively large to run the experiment. Finally, if you had 1B customers available, the minimum conversion rate you can detect is 4.003% (MDE = 3.3e − 5). With enlarged sample sizes you can really separate noise from signal.

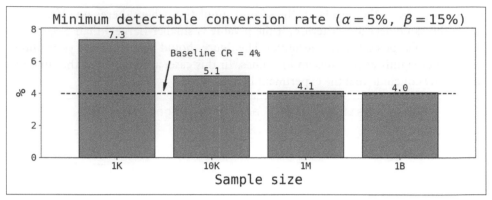

Figure 16-8. Conversion rate and MDE

Don't forget, the MDE refers to the *incremental* change in a metric caused by the treatment, which follows from the definition of the random variable you're doing inference on: $\hat{\theta} = \bar{Y}_B - \bar{Y}_A$.

As the example shows, once you fix the MDE, you can find the corresponding minimum detectable metric under the treatment, which would be:

$$\text{Minimum Detectable Metric} = \underbrace{\bar{Y}_A}_{\text{Baseline}} + MDE$$

Setting the MDE

At this point I hope I've convinced you that:

- Designing experiments must include statistical power and significance considerations that affect your sample size.

- With underpowered experiments, you might end up saying that a treatment had no effect, when the problem might actually be that you didn't use a large enough sample.

- You need to first set an MDE to find the minimum sample size for your experiment.

So how do you set the MDE in the first place? An important consideration is that *statistical* significance is different from *business* significance.

Going back to the previous example, even if you had 1B customers to include in your experiment (half gets the control and half the treatment), does it make business sense to run the test? What is the business impact of being able to detect a 3.3e − 5 incremental change? For most companies there's none, so even if the statistical properties are satisfied, from a business standpoint it doesn't make sense to go ahead with the experiment.[4]

You can use this type of reasoning to set a viable MDE with your stakeholders. For instance, for them it might make sense to find anything upwards of 4.1% conversion rates, so you must prepare to design a test for 1M customers. If you only have 10K available, you must discuss with them that you can only detect conversion rates upwards of 5% (MDE = 0.01). If everyone feels comfortable with such an estimate (25% increase relative to baseline), it makes sense to run the experiment.

Many times, your stakeholder won't be able to come up with an answer. If you have access to previous experiences, use those incremental changes (or an average) as your MDE to reverse engineer your sample size. Otherwise, use your business knowledge to come up with something that's sensible.

Hypotheses Backlog

A/B tests are only as good and informative as the hypotheses tested. I used to work at a company where the product teams routinely launched ill-designed experiments all the time. The lack of statistical robustness was not the most worrying aspect, however. Most of these experiments lacked well-founded hypotheses.

Having a hypotheses backlog is one critical aspect of developing a culture of experimentation inside a company. Ideally, this should include a *ranked* list of hypotheses that a team wants to test, along with the impacted metric and the arguments that support the effect. I'll discuss each of these now.

Metric

You won't be surprised to see that I start with the metric. Having a well-defined metric is a big part of the success of running an experiment. As discussed in Chapter 2, good metrics are measurable, relevant, timely, and actionable.

In A/B tests, the *closest* the metric is to the lever, the better, which tends to happen when the metric is both actionable and relevant. By this I mean that the lever moves the metric in a direct way, as opposed to a chain of effects that finally end up affecting the selected metric. Because of this, top-line KPIs are not great metrics to use when designing a test. As you might imagine, metrics decompositions can help you find the right metric for your A/B test.

---

4 Of course, if your company has one billion customers *to spare*, such a minuscule increase in conversion rates might generate substantial revenue, but bear with me for the purpose of this example.

Hypothesis

At a minimum, good hypotheses should be *causal* in the sense that you clearly state how and why the lever impacts the chosen metric.

The *how* refers to the directionality of the effect; for instance, the hypothesis *"if we lower the price by 1%, it's more likely that a customer will make a purchase"* clearly states that a price discount increases the conversion rate. This hypothesis still lacks the *why*, that is, an understanding of the mechanism behind the effect. The *why* is critical to assess the credibility of the hypothesis, and will also be used for ranking purposes.

Great hypotheses are also *risky*, not from a company perspective, but rather from the test designer perspective. Compare the following two statements that can easily follow the price discount hypothesis: *[the test matters because] the conversion rate will increase*, and *[the test matters because] the conversion rate will increase by 1.2 pp*. The former just provides directional guidance, and the latter quantifies the expected impact. Quantification provides important information that can be used to rank alternative hypotheses.

Ranking

It's good to understand that running experiments is costly for any organization. On the one hand, there are direct costs like the time, effort, and other resources used. But every time you interact with your customers, their perception of the company might change, possibly leading to customer churn or at least to lower future effectiveness (think customers flagging you as spam, thereby becoming unreachable). On the other hand, there's also the opportunity cost of launching tests that have a larger potential impact.

Once you consider the costs of launching tests, ranking different hypotheses becomes critical to guide their prioritization. A good practice is to share the hypotheses backlog across your organization so that different teams can participate and discuss the ranking and the other relevant information.

Governance of Experiments

If you make testing an integral part of your data-driven strategy, there will be a point where a governance framework needs to be implemented and formalized. As with data governance, I tend to stand on the more pragmatic side where, rather than trying to accomplish an exhaustive set of tasks, you aim to satisfy a minimalistic set of objectives (that are actually implementable).

Some objectives that might be important for your organization are:

Accountability
Experiments should have a clearly defined owner (usually a team) responsible for the results, intended or not, from the test.

Business safety
Reasonable guardrails should be implemented to guarantee that no one team's experiments can have a significant impact on the business. These guardrails should turn off the experiment if one or several KPIs exceed some predefined thresholds.

Customer and human centricity
Experiments that affect the behavior of humans, customers or not, should follow some minimal ethical standards that align well with the company's values.

Global versus local effectiveness
When several experiments are running at the same time, it's necessary to guarantee that treatment and control groups from different tests don't overlap. It might also be necessary to establish a policy on quarantine or resting periods so as to not affect the global effectiveness of business operations and other tests.

Knowledge incrementality
As key pieces for improved decision making, results from A/B tests should help grow and nurture a knowledge repository with positive and negative results.

Replicability and reproducibility
Any documentation and code used to design and analyze the results from experiments should be kept in a company-wide repository for later reproducibility.

Security
The technology stack used for running experiments at scale should adhere to the company's data security and data privacy policies.

Transparency and monitoring
Results should be made available as widely and timely as possible.

Key Takeaways

These are the key takeaways from this chapter:

A/B tests are powerful methods to improve an organization's decision-making capabilities.
You can think of A/B tests as performing local optimization of your organization's main metrics.

Tests should be designed to achieve a desired statistical power.

A/B tests should be designed taking into account the probabilities of a false positive or false negative. Statistical significances control the former, and power controls the latter. Underpowered experiments may lead you to incorrectly dismiss the true effect that your experiment has because of an insufficient sample size.

Quantifying an experiment's minimum detectable effect (MDE) should help you design tests with good statistical power.

Calculating the MDE is straightforward, and tells you the smallest incremental effect you can aspire to estimate for a given significance level and power, sample size, and variance of the outcome under consideration. For a given MDE, you can solve for the minimum sample size using the same formula.

Experimental governance.

As your organization becomes more mature and the number of simultaneous tests that are run scales, you will need to put in place a governance framework that allows you to achieve some minimal desirable objectives. I propose several that might suit your organization.

Further Reading

Howard Bloom's "The Core Analytics of Randomized Experiments for Social Research," *The SAGE Handbook of Social Research Methods*, 2008, available online (*https://oreil.ly/ZYG15*)), or his "Minimum Detectable Effects: A Simple Way to Report the Statistical Power of Experimental Designs," *Evaluation Review*, 19(5) (available online (*https://oreil.ly/QCxlC*)) should help you understand the derivations for the MDE formula. You can also check my notes to the Appendix (*https://oreil.ly/1S0Es*) of *Analytical Skills for AI and Data Science* (O'Reilly).

Part II of Guido Imbens and Donald Rubin, *Causal Inference for Statistics, Social, and Biomedical Sciences: An Introduction* (Cambridge University Press, 2015) discusses at great length many different aspects of statistical inference using randomization (A/B tests), such as model-based (Bayesian) inference, Fisher's exact *p* values, and Neyman's repeated sampling. Note, however, that they don't discuss design issues.

Ron Kohavi, Diane Tang, and Ya Xu's, *Trustworthy Online Controlled Experiments. A Practical Guide to A/B Testing* (Cambridge University Press, 2020) provides a book-length treatment of the many practical difficulties you may encounter when designing and running online tests at scale. A significantly shorter and condensed version can be found in Ron Kohavi and Roger Longbotham, "Online Controlled Experiments and A/B Tests," in D. Phung, G. I. Webb, and C. Sammut, eds., *Encyclopedia of Machine Learning and Data Science* (Springer, available online (*https://oreil.ly/DDRZd*))

Nicholas Larsen, et.al, "Statistical Challenges in Online Controlled Experiments: A Review of A/B Testing Methodology" (arXiv (*https://oreil.ly/R0uiR*), 2022) provides a recent survey on similar topics. For instance, I haven't discussed heterogenous treatment effects or SUTVA violations.

I found Sean Ellis and Morgan Brown's, *Hacking Growth: How Today's Fastest-Growing Companies Drive Breakout Success* (Currency, 2017) useful to design and implement successful hypotheses backlogs. While they focus exclusively on topics related to growth, the approach can be easily generalized.

## CHAPTER 17

Large Language Models and the Practice of Data Science

According to one estimate (*https://oreil.ly/2CoQ6*), almost four thousand jobs were lost from advances in AI in May 2023 in the US, representing almost 5% of all jobs lost in that month. Another report (*https://oreil.ly/xCO5d*) from a global investment bank estimates that AI could substitute 25% of all jobs, and OpenAI, one of the main players in the field, estimates (*https://oreil.ly/IhhLZ*) that almost 19% of all occupations have significant exposure, as measured by the fraction of tasks that could be impacted by AI. Some analysts (*https://oreil.ly/sq6AE*) claim that data science is itself *amenable* to be affected.

So how will large language models (LLMs) like GPT-4, PaLM2, or Llama 2 change the practice of data science? Will the *hard parts* presented in this book, or elsewhere, remain important for your professional development and career advance?

This chapter is quite different from the previous ones, as I won't discuss any techniques, but rather, I'll *speculate* on the potential short- and medium-term impact of AI on the practice of data science. I will also discuss whether this book's content might pass the test of time with the current disruption of AI.

The Current State of AI

AI is a broad field that encompasses many different techniques, methods, and approaches, but is generally associated with the use of very large neural networks and datasets. In the past few years, the pace of advance in the fields of image recognition and natural language processing has increased substantially, but it is the latter, with the latest releases of transformer-based LLMs such as OpenAI's GPT4 (*https://oreil.ly/*

tGzAm) and Google's Bard (*https://oreil.ly/ZWSZ4*), that created the current agitation and worries about their impact on the labor markets.

It's widely accepted that LLMs are great at performing natural language tasks, including text understanding and generation, summarization, translation, classification, and code generation. Interestingly, unexpected behavior has emerged as the size of the models reaches a certain threshold. This includes *few-shot learning* that allows the model to learn new tasks from a considerably small number of observations and *chain-of-thought reasoning* whereby the model solves a problem by splitting the argument into steps.

In a widely discussed paper on the impact of LLMs on the job market, Tyna Eloundou et al. (2023) look at specific tasks performed by different occupations and classify them into three groups, depending on their degree of *exposure* to productivity enhancements by AI (no exposure, direct exposure, or exposure through LLM-powered applications).[1] Among many other findings of interest, they show that some skills are more correlated with their measure of exposure. Figure 17-1 shows the correlation between basic skills and exposure from their analysis.[2] As you can see, *programming* has the strongest positive correlation with exposure and *science* the most negative correlation; this suggests that occupations that rely heavily on these skills are more or less exposed to impact, respectively.

What does this mean for data science? Just from the basic skills, you can hypothesize that some parts are highly exposed—most prominently programming—and others less so (specifically science and critical thinking). But it really depends on what you think a data scientist does. The truth is that data scientists perform a wide variety of tasks across companies, not only machine learning (ML) and programming.

1 They define "exposure as a measure of whether access to an LLM or LLM-powered system would reduce the time required for a human to perform a specific (detailed work activity) or complete a task by at least 50 percent."

2 In the plot I average the three estimates they report in their Table 5, so directionally speaking I'm capturing the intuition they wish to convey that some skills are more exposed to LLMs.

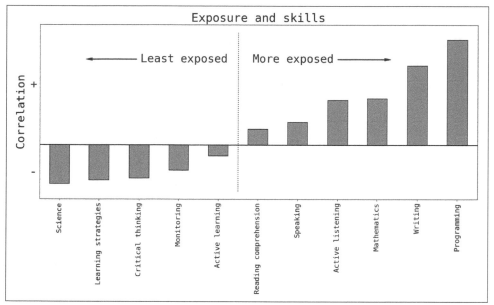

Figure 17-1. Skills positively and negatively correlated to exposure to impact (averages from Table 5 in Eloundou et al. (2023))

What Do Data Scientists Do?

To better understand data science's level of exposure to current AI, I'll now look at specific tasks done by practitioners in the workplace. For the purpose of the exercise, I'll use the list of tasks presented on O*Net (*https://oreil.ly/fCcZO*), which is commonly used in similar analyses; it's not perfect and it's certainly incomplete, but it still provides a useful benchmark.

Each task will be evaluated with respect to the four main basic skills used in data science: business knowledge, ML and statistics, programming, and soft skills. My sole aim is to provide a directionally correct assessment, so I'll use three possible levels (*low*, *medium*, and *high*), numerically coded as 0, 1, and 3, respectively, and denoted by x below.

For instance, I rank the task *analyze, manipulate, or process large sets of data using statistical software* as *high*, *low*, *high*, and *low* in business, ML, programming, and soft skills, respectively. To me, a deep knowledge of the business is required for *analyzing*, but other than that, this task relies heavily on *programming*. I repeat this same process for all tasks.[3]

---

3 You can find the actual rankings in the repo (*https://oreil.ly/dshp-repo*).

To get an exposure estimate, I use the following process:

1. Evaluate each task according to each of the four basic skills.
2. Compute exposure for each task using the following equation:

$$\text{Exposure} = \overbrace{0.2 \cdot x_B + 0.8 \cdot x_{ML} + 1 \cdot x_P - 0.2 \cdot x_S}^{\text{Basic skills}} - \underbrace{(0.2 \cdot x_B \cdot x_{ML} + 0.2 \cdot x_B \cdot x_P)}_{\text{Analytical skills}}$$

Here are my logic and assumptions behind this formulation:

- All basic skills can be learned by LLMs, at least to some extent. In terms of exposure, I rank them as Soft < Business < ML < Programming, hence the weights on the linear part. This order captures the intuition that, at least, in the short term, programming is more exposed than ML, and this in turn is more exposed than business knowledge, which in turn is more exposed than soft skills.
- I believe that ML and programming tasks that involve business knowledge require *analytical skills* and critical thinking that are going to be harder to develop until there is human-level or artificial general intelligence (AGI). Hence I include interaction terms that *lower* the exposure metric.

The importance of soft skills is worth discussing further: my take is that soft skills will remain extremely important in human-to-human interactions. But as far-fetched as it may sound, it's not hard to imagine a future state where AI completely replaces one of the humans in the interaction, and soft skills may become irrelevant.

The results are presented in Table 17-1. Out of the 15 tasks listed in the O*Net website, 40% are classified as skills with high exposure, 20% with medium exposure, and 40% have low exposure. Looking at the exposure metric for specific tasks, to me it looks directionally correct, and as expected, programming and ML tasks are more exposed, but the need for analytical skills reduces their overall exposure.

Each of the six tasks with a low exposure metric relies heavily on business knowledge, analytical skills, or soft skills. Analytical skills play a big role in my low exposure evaluation of all the skills where the data scientist *identifies or proposes* solutions.

On the other side of the spectrum, tasks that I evaluate as *highly exposed* are more easily automated with the current state of the art in AI. At this point in time, some of these still require an expert human in the loop, but this may not be the case in the near future.

Table 17-1. Data science tasks and exposure

Tasks	Exposure
Deliver oral or written presentations of the results of mathematical modeling and data analysis to management or other end users.	Low
Recommend data-driven solutions to key stakeholders.	Low
Identify business problems or management objectives that can be addressed through data analysis.	Low
Identify solutions to business problems, such as budgeting, staffing, and marketing decisions, using the results of data analysis.	Low
Propose solutions in engineering, the sciences, and other fields using mathematical theories and techniques.	Low
Read scientific articles, conference papers, or other sources of research to identify emerging analytic trends and technologies.	Low
Apply feature selection algorithms to models predicting outcomes of interest, such as sales, attrition, and healthcare use.	Medium
Analyze, manipulate, or process large sets of data using statistical software.	Medium
Design surveys, opinion polls, or other instruments to collect data.	Medium
Clean and manipulate raw data using statistical software.	High
Identify relationships and trends or any factors that could affect the results of research.	High
Test, validate, and reformulate models to ensure accurate prediction of outcomes of interest.	High
Apply sampling techniques to determine groups to be surveyed or use complete enumeration methods.	High
Compare models using statistical performance metrics, such as loss functions or proportion of explained variance.	High
Write new functions or applications in programming languages to conduct analyses.	High

Evolving the Data Scientist's Job Description

Taking the list of 15 tasks that data scientists commonly perform as a benchmark job description, and assuming the predictions about exposure are at least directionally correct, it is clear that the practice of data science will need to evolve as more powerful AIs continue to be deployed across companies.

At least with respect to programming, there seems to be a consensus that the current state of AI significantly improves the productivity of a developer. Tools like GitHub Copilot (*https://oreil.ly/wFQg7*) and Bard (*https://oreil.ly/a4RD9*) are becoming a standard, and there's every reason to believe that data scientists and data engineers are embracing these tools too. Some commentators have even talked about a 10x productivity enhancement (*https://oreil.ly/n-2WO*), and a recent survey (*https://oreil.ly/4k_AR*) finds that more than 90% of developers already use AI as a productivity tool.

At this point, however, it's clear that the current state of LLMs requires a human expert in the loop, both to prompt and guide the AI to the desired answer and to debug some errors that may arise and deal with any possible hallucinations (*https://oreil.ly/ZlGRN*). Also, as opposed to much work in pure software development, programming with data requires that the outputs make sense from a business standpoint, and at present this task requires a knowledgeable human.

But it's worth asking whether at some point in the not-so-distant future a business stakeholder will be able to interact directly with an AI, completely making redundant the job of a data scientist.

For instance, in many companies data practitioners take a business requirement from a stakeholder and code the necessary queries in SQL to generate reports or dashboards. This is one task that I think is highly exposed to AI and is thus likely to disappear from the future job description for data scientists.

So how will the future data science job description look? Assuming again that AGI hasn't been reached (otherwise, every occupation will have to be redefined), it seems to me there are two alternative long(ish)-term scenarios:[4]

- Nontechnical business stakeholders become *data-driven* and learn to ask questions from data and think analytically and scientifically about their business problems.
- Data scientists become business and analytically savvy and learn to make business decisions based on the evidence.

In the first scenario, the data science occupation disappears, and nontechnical business people undergo extensive retraining to acquire the skills needed to interact with an AI and answer business questions in a data-driven way (think of the earlier SQL example). In this case, the AI enhances the abilities of the business-savvy human.

In the second scenario, it's the business stakeholder that becomes redundant, and the data scientist uses the AI to perform the technical tasks and use their unique analytical skills and knowledge of the business.

Which scenario, if any, is most likely to emerge? My guess is that it depends on which skill is more costly to acquire, becoming data driven (and thinking scientifically about business questions) or becoming business-savvy. From my sole experience and observations in the previous decade where being data driven received a lot of attention, nontechnical business people have *not* made substantial improvements on the data-driven front. But neither have data scientists on the business savvy front. Maybe changes will finally take place when survival is at stake.

---

4 To be honest, the fast pace of advance makes it really hard to predict *when* things might end up happening.

Case Study: A/B Testing

I will use the case of A/B testing to explain some of these predictions and speculations. At its core, A/B testing requires two sets of skills:

Business
> Defining and prioritizing the hypotheses backlog to be tested, as well as the outcome metric to evaluate the experiment.

Technical
> Designing the experiment, randomizing and ensuring that assumptions for identification of causal effects are likely to be attained, and measuring the impact.

My guess is that in most companies, at present there is almost complete functional separation between data scientists and business stakeholders in these realms. But leaving aside the 1% of cases where highly specialized, frontier knowledge in the design and evaluation of experiments is required, my guess is that a big part of the technical details can very much be automated with current technologies (LLMs may just serve as a mediator). In my opinion, the really hard part of (nonfrontier) A/B testing is coming up with good metrics and good hypotheses to test.

Humans, enhanced by AI, should be able to take care of the vast majority of tests run at companies. Thanks to the vast corpus of data used to train the LLMs, I can also see humans using the AIs in the process of ideating hypotheses, but without a deep knowledge of how the world and the business works, and the underlying causal mechanisms, I just don't see AIs being critical in this arena.

To be sure, I don't see AIs running the technical side completely on their own. A knowledgeable human will guide the process. The question is who will that be, and what will their role be called.

Case Study: Data Cleansing

Data scientists spend quite a bit of time cleaning and transforming data to make it amenable for more valuable purposes. Again, I will assume that the *easy part* of the data cleansing process is the actual execution, using SQL or any other programming language like Python or R, as is commonly performed today.

The really hard part is making decisions that depend on critical business know-how. A typical example is whether you should convert null values to zeros or not. The answer is that in some cases it actually makes sense, but in others it doesn't. And it depends on the business setting. Another example is data quality, where you end up knowing that things are right because *they make sense* from a business standpoint.

Can a nontechnical business stakeholder take care of these decisions, enhanced by AIs that help out with the *easy* parts? I think that the answer is positive, but it may require some retraining, or at least some well-documented processes. Of course, it's not hard to imagine that in the future, these internal playbooks can be used in the training of a company's AI agents.

Case Study: Machine Learning

What about ML use cases? As a starting point, a data scientist decides which technique ought to be used for specific use cases. My guess is that with the current state of LLMs, AIs can easily help a nontechnical business stakeholder make this decision (because there are many discussions on the web on when to do what, and this is part of the corpus used to train the current family of LLMs). Put differently, I don't see humans having a comparative advantage in making this decision: again, putting aside the 1%-ish use cases where highly specialized talent is needed, all you need is a play-book that data scientists today learn on the job. Nonetheless, a critical aspect is understanding *why* one tool is better than another, and it's apparent that LLMs are still far from reaching this level of intelligence.

It is true that the best data scientists today specialize in the technical details of each predictive algorithm, and use this knowledge to fine-tune the models to make them more performant. For instance, it's very easy to train an out-of-the-box gradient boosting classifier, but it's harder to know which metaparameters to optimize to increase predictive performance. But the truth is that there already exist automated ML frameworks that take care of this. That's why I no longer think that this is a key skill that gives data scientists a comparative advantage over AI. Moreover, LLMs can, or will, be used to suggest an alternative course of action if needed (again, because of memorization/information retrieval of training data).

So what is the hard part about ML, where humans have a clear comparative advantage over LLMs? I believe it's coming up with hypotheses about the underlying causal mechanisms regarding why a set of features predicts a given outcome. Here, the *science* part of the job description is critical, and might give some advantage to data scientists over nontechnical business people in the future.

LLMs and This Book

This book presented techniques aimed at helping you become a more productive data scientist. Naturally, some of these techniques are more or less exposed to AI, depending on the combination of basic underlying skills required to perform them.

In Table 17-2 I show my subjective assessment of each chapter's exposure, using the exact same methodology as before. Again, my aim is to be directionally correct only, and the results look reasonable to me: chapters in the first part rely more on business knowledge and soft skills, and are less intensive in programming skills and ML or statistics knowledge, so they are less exposed. The second part deals more with ML and statistics, hence the higher exposure.

Table 17-2. Book chapters sorted by exposure

Chapter	Main lesson	Exposure
1. So What	How to measure the impact of your team	Low
7. Narratives	How to build narratives before and after creating a project	Low
6. Lift	Technique to see differences across groups	Low
2. Metrics Design	Find better metrics to action	Low
3. Growth Decompositions	Understand what's happening with the business	Low
4. 2×2 Designs	Simplify to understand complex problems	Low
5. Business Cases	How to measure the impact of specific projects	Low
8. Datavis	Extract knowledge and convey a key messages with datavis	Medium
15. Incrementality	Understand the basics of causality	Medium
16. A/B Tests	Design of experiments	Medium
10. Linear Regression	Strengthen your intuition of how ML algorithms work	Medium
13. Storytelling ML	Storytelling to create features and to interpret results	High
14. Predictions to Decisions	Making decisions from ML	High
9. Simulation and Boot	Tools to deepen your understanding of ML algorithms	High
11. Data Leakage	Identify and correct data leakage	High
12. Productionizing Models	Minimal framework to deploy in production	High

What does this say? Let's imagine that tasks are either *not exposed* or *completely exposed*; the former means that LLMs are of no value, and the latter means that AIs can do the tasks by themselves. The truth is that most tasks across occupations are somewhere in the middle, but let's dismiss this for now. In this extreme world, you should be investing in skills of the former type, as they make you special, relative to LLMs.

The main point of this exercise is that some of the skills learned in the book are worth more of your time and effort to develop, given the current capabilities of LLMs. Note that I'm not saying that you should not invest in becoming a great programmer or an expert in ML or stats. Rather, at least for programming, the rise of LLMs has made this skill less valuable for you as a data scientist. For ML and stats it's still too early to call.

My predictions are, of course, to be taken with a grain of salt, but I do think that it's certain that the future of the practice of data science may follow a path where in the short term, a data scientist's productivity is augmented by the sole power of LLMs to generate high-quality code, assisted by knowledgeable humans. The longer term is much more uncertain, and it's not unreasonable that the data science practice will be completely redesigned or even cease to exist, as discussed previously.

Key Takeaways

These are the key takeaways from this chapter:

LLMs are changing the workplace.
2023 will likely be remembered as the first year where AI started to create a measurable impact on the workforce and the labor markets.

Data science is being impacted as we speak.
Similar to software developers, an immediate impact of AI on the practice of data science is on programming productivity.

But many other tasks commonly performed by data scientists are also exposed to AI.
I analyze 15 tasks listed by O*Net, and find that the level of exposure is high for around 40% of the tasks, and medium for 20% of the tasks. Tasks that rely more heavily on programming are naturally more exposed, but I assume that machine learning and statistics are also going to be impacted in the medium term. Business knowledge and soft skills are assumed to be less exposed.

Changes in the job description for data science.
My best guess is that the practice of data science will change in the near future, putting less weight on programming and ML skills, and more emphasis on analytical skills, causality, and business knowledge.

Further Reading

Suggested readings on this topic are very likely going to become outdated very quickly, given the speed of advance of the field. That said, here are some articles that have guided my understanding of the state of the field.

Tyna Eloundou et al., "GPTs are GPTs: An Early Look at the Labor Market Impact Potential of Large Language Models," March 2023, retrieved from arXiv (*https://oreil.ly/lDoUs*). This paper provides exposure to AI estimates across occupations. I don't use the same methodology to quantify the level of exposure of data science, so this paper is certainly worth a read if you want to come up with alternative scenarios for the practice.

Sebastien Bubeck et al., "Sparks of Artificial General Intelligence: Early Experiments with GPT-4," April 2023, retrieved from arXiv (*https://oreil.ly/aN_xl*). This paper started an interesting debate on whether we are close to achieving AGI. They argue that in the future this family of LLMs will most likely be labelled as proto-AGI. Note that many leading researchers, most notably Yann LeCun (*https://oreil.ly/rj8tu*) (see also here (*https://oreil.ly/2x2KQ*)), believe that autoregressive models can't lead to AGI.

Ali Borji, "A Categorical Archive of ChatGPT Failures," April 2023, retrieved from arXiv (*https://oreil.ly/Q9K0V*). This paper is constantly updated and shows how things can go awry with the current state of AIs.

Grégoire Mialon et al., "Augmented Language Models: A Survey," February 2023, retrieved from arXiv (*https://oreil.ly/o_WWd*). Even if LLMs haven't attained human-level general intelligence, there are ways to improve their ability to reason or use external tools, spanning an even larger spectrum of use cases.

The following two papers discuss emergent capabilities in LLMs as their size continues to increase:

Jason Wei et al., "Emergent Abilities of Large Language Models," October 2022, retrieved from arXiv (*https://oreil.ly/ZcWNn*).

Rylan Schaeffer et al., "Are Emergent Abilities of Large Language Models a Mirage?," May 2023, retrieved from arXiv (*https://oreil.ly/CEqdI*).

Index

About the Author

Daniel Vaughan is currently a freelance data scientist and ML/AI practitioner and strategist. He is the author of *Analytical Skills for AI and Data Science* (O'Reilly, 2020). With more than 15 years of experience developing machine learning models and more than eight years leading data science teams, he is passionate about finding ways to create value through data science and in developing young talent. He holds a PhD in economics from NYU (2011). In his free time he enjoys running, walking his dogs around Mexico City, reading, and playing music.

Colophon

The animal on the cover of *Data Science: The Hard Parts* is a zebrafish (*Danio rerio*). The zebrafish is a freshwater fish in the minnow family and is native to South Asia. They are named for their five horizontal blue stripes running along their sides to the end of the tailfin. Males have gold stripes between the blue stripes, while females have silver stripes instead of gold. In the wild, zebrafish typically reach up to 1.5 inches long and live two to three years. They generally live in shallow water, including streams, ponds, and rice paddies.

Zebrafish are popular aquarium fish because, in addition to their vivid colors, they are easy to care for and easy to breed. Their eggs hatch in two or three days, and they reach maturity in three to four months. They are also popular in scientific research as vertebrate model organisms, partly because their transparent eggs and larvae make it easy to observe their development.

Because of their abundance in their natural habitat, zebrafish are considered a species of least concern. Many of the animals on O'Reilly covers are endangered; all of them are important to the world.

The cover illustration is by Karen Montgomery. The cover fonts are Gilroy Semibold and Guardian Sans. The text font is Adobe Minion Pro; the heading font is Adobe Myriad Condensed; and the code font is Dalton Maag's Ubuntu Mono.